Beyond the Neighborhood Unit

Residential Environments and Public Policy

ENVIRONMENT, DEVELOPMENT, AND PUBLIC POLICY
A series of volumes under the general editorship of
Lawrence Susskind, *Massachusetts Institute of Technology, Cambridge, Massachusetts*

ENVIRONMENTAL POLICY AND PLANNING
Series Editor:
Lawrence Susskind, *Massachusetts Institute of Technology, Cambridge, Massachusetts*

CAN REGULATION WORK?
Paul A. Sabatier and Daniel A. Mazmanian

PATERNALISM, CONFLICT, AND COPRODUCTION
Learning from Citizen Action and Citizen Participation in Western Europe
Lawrence Susskind and Michael Elliott

BEYOND THE NEIGHBOHOOD UNIT
Residential Environments and Public Policy
Tridib Banerjee and William C. Baer

RESOLVING DEVELOPMENT DISPUTES THROUGH NEGOTIATIONS
Timothy J. Sullivan

ENVIRONMENTAL DISPUTE RESOLUTION
Lawrence S. Bacow and Michael Wheeler

Other subseries:

CITIES AND DEVELOPMENT
Series Editor:
Lloyd Rodwin, *Massachusetts Institute of Technology, Cambridge, Massachusetts*

PUBLIC POLICY AND SOCIAL SERVICES
Series Editor:
Gary Marx, *Massachusetts Institute of Technology, Cambridge, Massachusetts*

Beyond the Neighborhood Unit

Residential Environments and Public Policy

Tridib Banerjee and
William C. Baer

University of Southern California
Los Angeles, California

Plenum Press • New York and London

Library of Congress Cataloging in Publication Data

Banerjee, Tridib, 1940–
Beyond the neighborhood unit.

(Environment, development, and public policy. Environmental policy and planning)
Bibliography: p.
Includes index.
1. Urban policy—United States. 2. Urban policy—California—Los Angeles—Case studies. 3. Neighborhood—United States. 4. Neighborhood—California—Los Angeles—Case studies. 5. City planning—United States. 6. Los Angeles (Calif.)—City planning—Case studies. I. Baer, William C., date–
II. Title. III. Series.
HT167.B36 1984 307′.3362′0973 84-11619
ISBN 0-306-41555-0

A Division of Plenum Publishing Corporation
233 Spring Street, New York, N.Y. 10013

Printed in the United States of America

To those four hundred and seventy-five Los Angeles
area residents, who, by gracefully agreeing to share
their residential life experiences with us, helped
us to formulate our own final thoughts about the
meaning and significance of the residential environment
in people's lives.

Preface

Much of the research on which this book is based was funded almost a decade ago by separate grants from two different agencies of the U.S. Public Health Service, of the then still consolidated Department of Health, Education, and Welfare. The first grant was from the Bureau of Community Environmental Management (Public Health Service Research Grant 1-RO1 EM 0049-02), and the second from the Center for Studies of Metropolitan Problems of the National Institute of Mental Health (Public Health Service Grant RO1 MH 24904-02). These separate grants were necessary because of budget cuts that truncated our original effort. We were fortunate to receive subsequent assistance from NIMH to conclude the research, as it is doubtful that a project of the scope and intent of our effort—even as completed in abbreviated form—will be funded in the 1980s.

The original intent of this project, as formulated by our colleagues Ira Robinson and Alan Kreditor, and as conceptualized earlier by their predecessors—members of an advisory committee of planners and social scientists appointed by the American Public Health Association (APHA)—was to rewrite *Planning the Neighborhood,* APHA's recommended standards for residential design. In particular, it was proposed that the new study take the point of view of the user in terms of residential standards. Hitherto, the private sector had dominated these considerations (i.e., the designer's predilections, the requirements of builders and material suppliers, and lenders' needs for mortgage security). Even where the public interest intruded (in matters of health, safety, and welfare), the standards had focused on the physical and had concentrated on prevention. The reformulation of these standards would insert the concerns of the user by focusing on their preferences and would promote physical and mental well-being by looking at the state of satisfaction of the users.

The research was to have three stages. In the first stage, various groups of people (selected by income class, race/ethnicity, and stage in the family cycle) would be interviewed to elicit their conceptions of an appropriate residential environment. During the second stage, these data would be analyzed and converted into revised planning standards that would meet the needs of each of these groups. The final stage would consist of a "game" composed of all of the major actors (environmental designers, builders and real-estate developers, lenders, and city officials) in which these tentative standards would be tested for utility and effectiveness. Based on the results of this "game," the tentative standards would be revised before final publication.

However, as explained previously, budgetary constraints meant that the larger goal, as well as the scope of the research required, had to be modified considerably after the project was started. The sample size was reduced; only a portion of the second stage (analysis of the data) was completed; and the third stage was eliminated entirely.

In one respect, this was unfortunate because the project showed great promise in terms of advancing the state-of-the-art of standard formulation. In another way, however, our partial effort provided some unanticipated benefits. Some perspectives were gained on the original problem that had not been appreciated when the project was first formulated.

For one, a subsidiary purpose of the original project was to devise some new methodological approaches so that local planners could adopt standards suitable to their individual areas based on special needs of the people. Thus, we pursued a research approach strongly recommended by our panel of technical advisers: an orientation toward the activities conducted in residential areas rather than just the physical layout and facilities provided there. Moreover, we attempted to elicit people's trade-off preferences (how much of one attribute they would sacrifice to achieve an increment in another). Both approaches were new to the social sciences, and in both areas, we have concluded that substantially more methodological work is necessary before any policy conclusions can be drawn from present findings.

For another, we now believe that insufficient concern had been placed on matters of equity in the allocation of resources to meet the standards. Yet, these issues have become the focus of the current debate on the patterns and causes of inequality in public service delivery. And we have come to agree that it is more important to address these fundamental conceptual questions first, before writing new standards.

Also, because the initial formulation of this research project focused on standards, the importance of the design paradigm that guides the planning and design of residential environments was not sufficiently appreciated or even adequately conceptualized. Although the premise of the neighborhood unit concept was very much under scrutiny from the outset, the immediate concern then was whether it could continue to serve as the organizing model for future residential standards. But although its validity was tacitly rejected, the possibility of developing alternative design paradigms was never considered. Furthermore, in that early stage of research, we were still guided by the notion that residential quality is solely a function of public policy at the neighborhood level; we had yet to define the problem in the framework of citywide design and public policy, which we propose in this book.

We believe, therefore, that even though a truncated version of the original undertaking is presented here, it should nonetheless prove valuable for several reasons. First, as stated at the outset, even the limited scope of the work that we do present here is unlikely to be funded during the 1980s, or perhaps even in the foreseeable future. It is an expensive undertaking.

Moreover, the data have an important regional orientation. Most social-science findings pertaining to residential development have a northern and eastern bias, with regard not only to physical development but also to the lifestyle associated with that development. This is not a deliberate bias; it merely reflects the location of most research institutions. Nevertheless, with the trend toward sprawling, low-density urban form, built with the automobile as the chief means of transportation very much in mind, and with the trend of population migration and development toward the South and the Southwest, and with Southern California as a pace setter in suburban lifestyles, it is appropriate and pertinent to balance past residential research with data from the Los Angeles region. Additionally, if we are to devise policy with an eye toward future problems of urban development, rather than exclusively toward existing or past problems, our research findings should have continued relevance for the coming decade.

Finally, we believe that, although our research project has a long history, the nature of residential experiences for different social groups has not changed significantly during this period. We believe strongly that the fundamental patterns of inequities and inequalities in the residential experiences of different social groups have remained essentially the same despite small and expected shifts in consumption patterns and lifestyles. Indeed, the current recession, cutbacks in government spending in social programs, and the "housing crunch" may even have exacerbated the patterns of inequalities that we report from our study. What has changed significantly during these intervening years, however, is our perception of the problem. In part, this change was a function of our own development and the general enrichment of intellectual content in this particular field and in the field of planning in general. But more significantly, it was the initial findings of the research that shaped our views and influenced the arguments that we present.

Acknowledgments

A special kind of acknowledgment must be noted at the outset, for we did not originate the project that we report here even though we finished it. Rather, as we relate in the Preface, it was originally conceived by Ira M. Robinson (now Professor of Environmental Design at the University of Calgary) and Alan Kreditor (currently Dean of the School of Urban and Regional Planning at the University of Southern California), who hired us to work on their initial project. Although they may not want to acknowledge much relationship with our finished product, we wish to note here our gratitude for their early faith in our efforts, and for the rare opportunity to work on a truly interdisciplinary project.

The conceptual foundation of this project was first laid in a 1969 conference organized by Ira Robinson, in which five position papers on various aspects of planning, designing, and managing the residential environment were presented by Francis Hendricks and Malcom McNair; Barclay Jones and his associates; Alan Kreditor; Anatole Solow and his colleagues; and Maurice Van Arsdol. The content of these papers and their reviews, made by such notable social scientists as Janet Abu-Lughod, Donald Foley, Robert Gutman, Irving Hoch, and Paul Niebanck, have always been an important intellectual resource for us to draw on from time to time during the life of the project, especially whenever we felt that we might be losing sight of the larger purpose of this study.

The overall research objectives and the focus of the inquiry were further refined and expounded upon by another group of experts who met in Los Angeles in 1971, just before we actually joined the project staff. The content of the discussions among these experts, as summarized in various minutes and memos, served as important background for our research design. The direction of the survey research was significantly influenced by the opinion of these experts, and here, we would like to thank Ali Banuazzi, Leland Burns, Daniel Carson, Stuart Chapin, Ido de Groot, Steve Frankel, Marc Fried, Chester Hartman, Stanley

Kasl, William Michelson, George Peterson, Kermit Schooler, David Stea, Gary Winkel, and Myer Wolfe, who constituted different expert panels.

The final research design was worked out in its early days by a team of planners and social scientists, of which we were members. Here, we wish to give our thanks and to pay tribute to the members, each of whom frequently felt that he or she was sacrificing disciplinary purity for some will-o'-the-wisp of policy relevance that none of us could well define. Nonetheless, we have found during the later stages of the project that the study design was well conceived for most of our purposes, and that the innovations we undertook back then hold up well today. Kay Bickson, Keith Corrigal, Peter Flachsbart, Joyce Herman, Renee Gould, Constance Perin, Marsha Rood, Gary Schalman, and Ed Sadalla were our colleagues on the team in those days; Virginia Clark provided help in our sampling design; and Deborah Hansler advised us on the survey methods. We trust they will not be too disappointed in our efforts to see many of their ideas through to completion, even if not performed to their standards. Peter Gordon, still our colleague at USC, also offered ideas and suggestions, although we fear that we did not pay close enough heed to some of his early warnings, which we fully appreciated only later. We thank Kay Duckworth and her associates for the screening interviews. A number of our students at the University of Southern California in those early days did the interviewing, a service for which we are also grateful.

In the second stage of the project, the diligent efforts of Frank Wein, Syyed Mahmood, Marc Melinkoff, Jim Barber, and Daniel Green rounded out the interviews and shaped the data into some kind of coherent form for analysis. TELACU, a planning organization in East Los Angeles, provided an invaluable service at this stage in conducting the interviews of some of our Hispanic respondents in Spanish. Steve Pierce and Charles Noval were especially helpful in our preliminary research analysis, as was Jill Sterrett, who provided research assistance on our first draft. We note also Richard Fujikawa's contribution in completing the final manuscript.

There was a hiatus following the first draft, during which we digested the difference between what we had written and what we thought we should have written. Of those who had an early glimpse of our first draft, the late Kevin Lynch and the late Donald Appleyard (who was also a member of the earlier expert panel) were the first to point out that it needed work. (In the years to follow, Donald periodically reminded us that we must get that book finished. His warm encouragement meant a lot to us, and we would have liked very much for him to see the final product.) Here, we also wish to express our appreciation for Frankee Banerjee's timely critique, which finally got us away from the trees so that we could view the forest again. Whatever faults the final version has, they were worse before, and we think there were more of them.

The revised draft went through innumerable versions, but in particular, we wish to thank Shirley Rock, for typing several drafts and for her patience with our last-minute decisions and modifications, and Marilyn Ellis, for her assistance in the final weeks of finishing the manuscript.

We also owe a debt of thanks to our colleagues in the School of Urban and Regional Planning and elsewhere who never smirked in our presence when we talked year after year about finishing our book. Their politeness made our own frustrations easier to bear.

We owe a special note of thanks to Lawrence Susskind, the Plenum series editor, for his advice and constructive suggestions, and to Gary Hack and Victor Regnier, who read the final draft and offered some additional insights and ideas to improve the final version. Marjorie Cappellari's editing was timely and enormously helpful. In addition to the grants received from the Public Health Service (which are acknowledged in the Preface), we must also note that a United Parcel Service Foundation sabbatical grant through Stanford University helped to offset additional costs of data analysis long after the main grant expired. We would also like to thank the administration of the USC School of Urban and Regional Planning for numerous support services and in-kind assistance.

Finally, we wish to express our deepest thanks to our wives, Frankee Banerjee and Susie Baer. Lord knows what they said to each other about our slow progress, but to us, they were the quintessence of support and understanding. To our children, too, we owe at least an acknowledgment for the time that we thought we should spend with them but could not because of the book.

Contents

1

Introduction

This book is about the residential life experiences of different social groups in a metropolitan area and what they mean for environmental design and public policy. Residential environment is an important domain of quality-of-life experiences; good residential environments enhance life satisfaction and the individual's overall sense of well-being. Environmental design and public policy are both important in shaping the quality of residential environments. It follows, therefore, that both designers and public policymakers should be apprised of how people perceive, use, image, and value their environment and what they need and expect from it. A good environment cannot be built without an understanding of its social and human purposes.

This has been the motive for our research. We contend (as do other scholars)[1] that the environmental design and the public policies that have shaped much of the existing residential settings traditionally have been formulated without benefit of empirical research. However, we believe that this omission is due no longer to the lack of empirical research *per se,* but to the questions asked and the form in which that research was carried out as compared with the problems that designers typically face. A gap exists that is rarely bridged. Thus, a secondary purpose of the research that led to this book was to conduct research of use to environmental designers as well as social scientists. We do this not only by reporting research findings but also by discussing their implications for the future planning and design of residential environments.

The core of this research is a popular residential-design paradigm[2]: the *neighborhood unit.* This paradigm has been used widely as a design concept both in the United States and in other parts of the world, although it has no scientific foundation; and it has endured—even prospered—for fifty years, despite early, frequent, and telling criticism. Not only has this concept contributed significantly to the existing quality of the residential life experience, it also encapsulates the old debate between physical and social determinism, the differences between policy analysis and policy formulation, and in a fundamental sense, the nexus and dilemmas between art and science.

The fact that the neighborhood unit has been extensively used as well as extensively criticized in both design practice and design literature, and that it has, as well, been a focus of research in the social sciences, makes it a convenient core around which to base our study. *However, we must make clear at the outset that the major focus of this work is not the "neighborhood unit" per se, but what we term the "residential area."* While the *neighborhood unit* concept is used to focus research on a well-established planning paradigm, we use the *residential area* concept in the main to express our results. We believe that the term *neighborhood* is too restrictive in orientation to capture fully the perceptions and evaluations of our respondents, and that it is too "loaded" in value orientation among researchers (either pro or con) to allow dispassionate discussion. The meaning of *residential area,* on the other hand, has more neutral connotations. Its precise meaning will become clear as we show how our respondents defined their "residential area" in the course of the survey.

THE NEIGHBORHOOD UNIT AS A RESEARCH AND DESIGN PROBLEM

The concept of the neighborhood unit was originally formulated in the 1920s. It offered in concrete terms a model layout of a neighborhood of a specified population size, with specific prescriptions for the physical arrangement of residences, streets, and supporting facilities. Based on the then-popular notion of separation of land uses and the segregation of vehicular and pedestrian traffic, it emphasized boundaries and an inwardly focused core. Thus, it was celluar and was relatively self-contained, to be used in building-block fashion to construct a larger urban realm of neighborhoods.

The prestigious American Public Health Association adopted it in the late 1940s as the basis for formulating "healthful and hygienic" standards for planning, designing, and managing the residential environment. Subsequently, it was adopted, modified, and institutionalized by various professional organizations and public agencies. It was incorporated into many local planning manuals and zoning ordinances in the United States, and it was used in New Town development in other parts of the world. Planners have embraced its purpose of creating a sense of community; public agencies have adopted its purpose of protecting (and promoting) the public health, safety, and welfare; private developers and lending institutions have sought its protection of property values and investment decisions. For more than fifty years, it has been virtually the sole basis for formally organizing residential space. Even when not specifically invoked, its premises and constructs have guided residential planning and design. Its credentials are impeccable, its position preeminent, and its use ubiquitous. But is the paradigm unassailable? (See Figures 1.1–1.3.)

Since the late 1940s, the concept has been under strong attack. Some planners have questioned its unintended consequences, and many social scientists have questioned its premises. Thus, it has been argued that the concept would prove socially divisive—that it encouraged and fomented the very segregation that society has increasingly rejected; that it emphasized the physical environment as the prime determinant of residential quality of life when, in fact, the social environment was more salient; that it was an increasingly obsolete

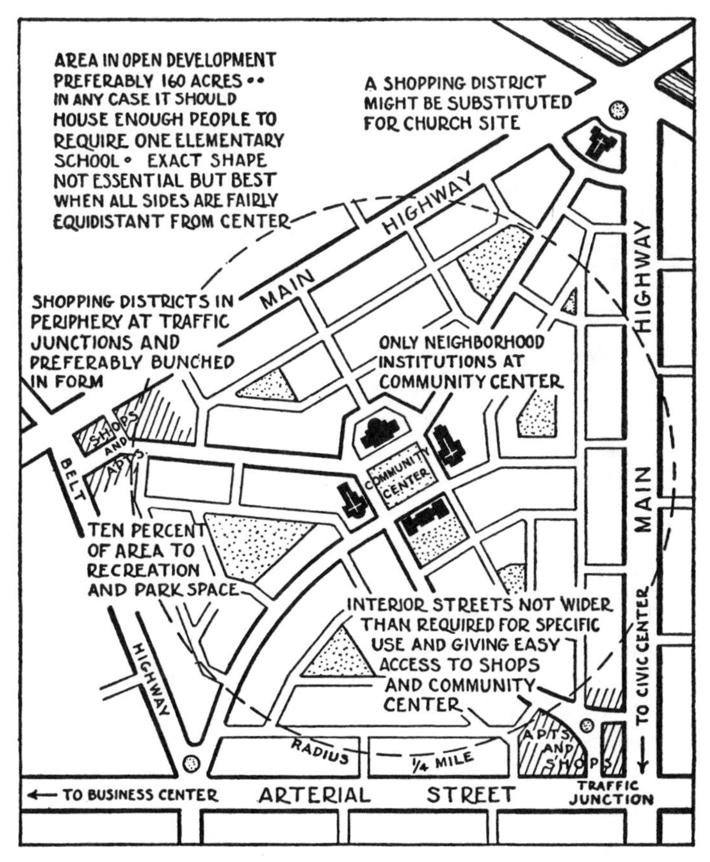

FIGURE 1.1. Graphic representation of the neighborhood unit principles as originally conceived for the New York Regional Plan. (Source: Committee on Regional Plan of New York and Its Environs. *Regional Survey of New York and Its Environs,* vol. vii, New York, 1929. Reproduced with permission from Regional Plan Association.)

contrivance geared to the needs of yesterday's rural migrant in need of a sheltered villagelike existence, even though the city-dweller has become increasingly urbane, using a nonplace urban realm as a satisfying environment.

Many of these early criticisms were conjectural, but of late, a wealth of

FIGURE 1.2. Application of neighborhood unit principles in the general plan for the Radburn New Town. (Source: C. S. Stein, *Toward New Towns for America.* Cambridge: M.I.T. Press, 1957. Copyright 1957 by Clarence S. Stein. Reproduced with permission from M.I.T. Press.)

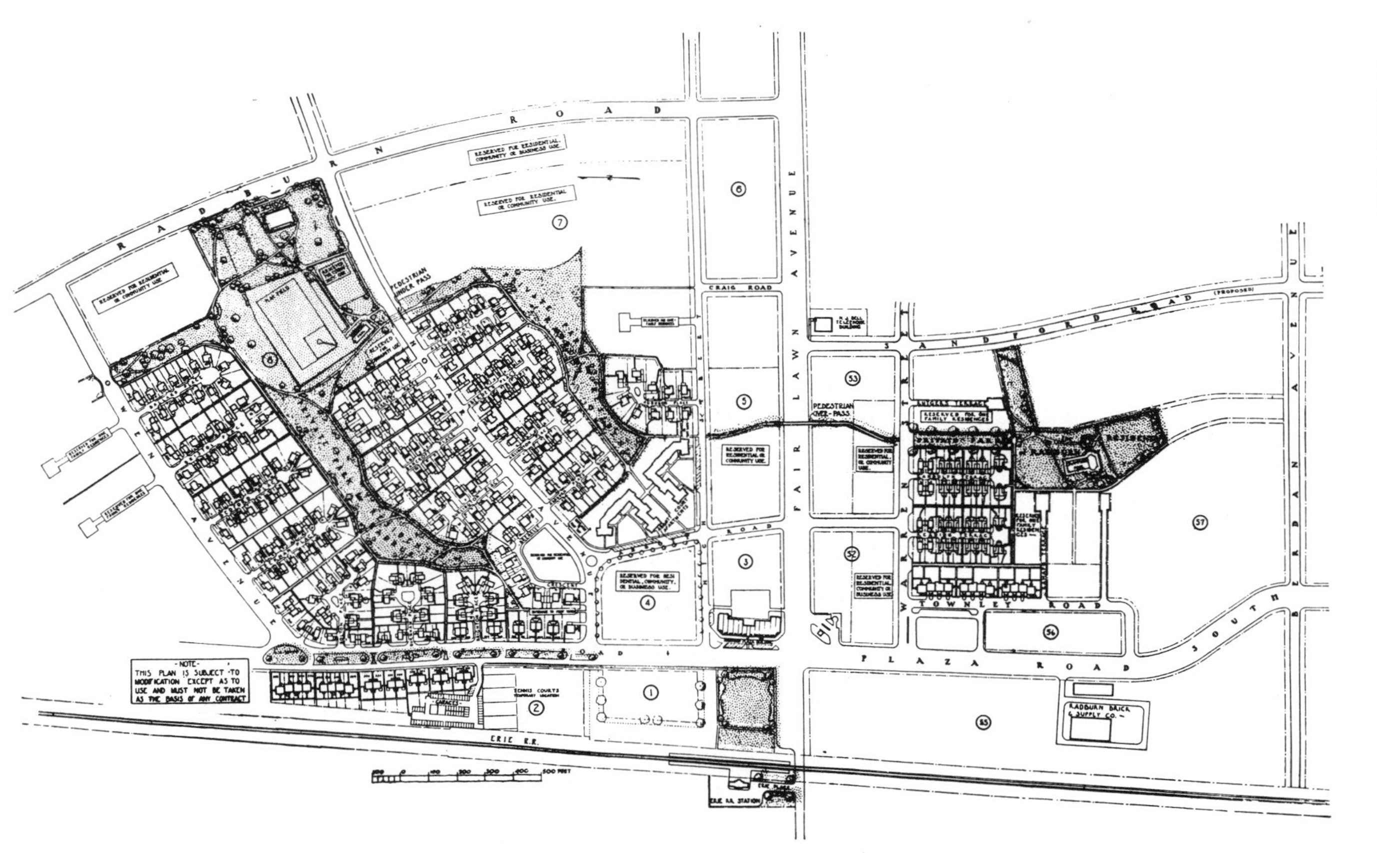

FIGURE 1.3. Detailed plan of a residential district showing the neighborhood unit principles in conjunction with the superblock idea. (Source: C. S. Stein, *Toward New Towns for America*. Cambridge: M.I.T. Press, 1957. Copyright 1957 by Clarence S. Stein. Reproduced with permission from M.I.T. Press.)

studies has been amassed, both in the United States and abroad, that temper the more extreme criticisms of the concept. These suggest that although the neighborhood concept was helpful at times in providing a sense of place to the inhabitants, people's lives are considerably more complex than had been earlier conceived, warranting perhaps a greater variety than could be provided by the neighborhood unit.

But despite this argument and despite all the research, no alternative paradigms that respond to these shortcomings have been forthcoming. As a rule, any profession has competing theories, alternative hypotheses, or differing schools of thought about how to solve a problem. But in the area of residential design, we have found no alternatives. One either believes in the neighborhood unit, is an agnostic, or is an atheist. There are no rival gods—as yet!

How was the neighborhood unit derived? The answer points up one of the fundamental difficulties in devising scientific research that meets designers' needs. The formulation was not the culmination of years of painstaking testing of a hypothesis by myriad individuals trained in science. Rather, it sprang full blown from the mind of one man, Clarence Perry, and even then, it was not the result of his having carefully sifted through research findings and having put together a construct derived from evidence on residential behavior gathered from a variety of sources. Instead, it was created as a whole, modified from earlier design experiences and reflecting intellectual ideas of Perry's era. Rather than presenting a scientific theory to be subjected to rigorous questioning and testing by others, Perry presented a design pattern to be utilized by others to the extent that they believed in it intuitively.

The difference lies in the underlying epistemology of the design arts as compared with the sciences. For scientists, the operative word is *analysis*—the separation of a whole into its component parts. For designers, the operative word is *synthesis*—the composition or combination of parts or elements so as to form a whole. The former understand the whole only after knowing about each of the parts in detail; the latter understand each part only after reference to the whole. This formulation is overstated to make a point. Science is not oblivious to synthesis; design is not unaware of analysis. But the guiding force is different in each case.

The problem with which we are concerned is translating from the expertise of one to the expertise of the other. On the one hand, we find that social science findings, although obviously pertinent to designers, are presented in a form that is of little value in designers' subsequent practice. Too often, social scientists leave their findings scattered about without any guide to how to put them together into a whole. The designers' program, or "brief," on the other hand, is so intricately woven that social scientists frequently fail to comprehend the entirety of the problem that designers face. Instead, they investigate only a part, not realizing that it is the whole that they must work with. The designer is trained in the art of "satisficing" or formulating "second-best" solutions (where no one part is entirely perfect or "true" but where each fits nicely with the others) that allow completion of the project. Scientists, however, have not been trained to accept the validity of these principles and will leave a theory only partially constructed, awaiting the day when someone else will figure out the missing piece in its "true" (optimum) formulation. In short, the scientists strives for absolute truths for each part of a theory (even if some parts must remain missing for a while), whereas the designer accepts partial truths if enough of the pieces are present to allow some kind of whole to be produced. The difference is in the purpose that each serves. A completed science composed of half-truths is of as little value to the scientist as a half-finished building composed of perfectly designed parts is to the designer.

The success of Perry's formulation lay not in its presentation of a set of scientifically derived findings but in its offering to designers for the first time a means of ordering and organizing—at least conceptually, if not in reality—the city into subareas that seemed to satisfy a number of social, administrative, and service requirements for satisfactory urban existence. Moreover, it organized a confluence of social values that were considered important in the early twentieth century into a physical form comprehensible to and usable by environmental designers.

Environmental design is, by and large, paradigmatic.[3] It tends to use preconceived models or patterns of a whole for its basis rather than drawing directly (and *de novo*) each time from theory or empirical findings. Thus, design consists largely of variations on an accepted stylistic theme rather than of an entirely original conception devised from scratch for each design problem. Certainly, the history of city design demonstrates this proclivity of professional designers (Lynch, 1981), and some design theorists argue that this practice is beneficial, that paradigms are the very stuff from which efficient designs are made (Alexander *et al.*, 1977). But these design paradigms lack science's perpetual self-questioning. They have no tradition of continual testing of their validity. As design is for people, and people and society change over time, there is always the risk that even if a design paradigm is correct for one age, it may be incorrect for another. In design, competing paradigms have traditionally served as a test of sorts, but as we have seen with the neighborhood unit, there is no alternative. Thus, despite its obsolescence and inappositeness, the neighborhood unit persists.[4]

We have already alluded to the numerous criticisms of the neighborhood unit, and we describe these at greater length in the next chapter. These criticisms question the basis on which the concept is constructed. But we have also suggested that whatever the merit of the concept at the time of its formulation in the 1920s, times have changed, society's values have shifted, and, accordingly, the relevance of the concept has waned. For instance, the social consensus in the

1920s dealt with public health and hygiene, and with "social betterment" and physical design, the latter being not well understood at the time, and it was linked with the now-discredited notion of physical determinism. Rural-urban migration, citizenship, and, in general, the heritage of the recent Reform Movement, which expressed itself in a variety of policy areas (Lubove, 1962), were also of concern in the 1920s, but they are of much less importance at present. Furthermore, the neighborhood unit paradigm assumed a cultural homogeneity that, if it ever existed, certainly is less apparent and hence less applicable today. Ethnic and cultural pluralism is currently a widely accepted tenet in our nation's political beliefs, and this pluralism may express itself in a variety of ways, including everyday residential life. Thus, there could be differences in needs and preferences of different racial and ethnic groups, and as the 1980 census data show, these groups are becoming a larger percentage of the metropolitan population.

The days when the family with children—who epitomized the neighborhood unit residents—held complete sway in determining housing-market demands are over. Singles, "mingles," professional couples without children, elderly people who insist on living apart from their children or relatives, gays, and others advocating alternative lifestyles all compose a larger proportion of the housing market than in years past, and each group has values and hence residential needs that are somewhat at variance with the family values embodied in the original neighborhood-unit concept.

Finally, there has been a recent questioning of the assumptions underlying current public policies that affect the quality of residential environments. Many of these policies are based on design standards promulgated by professional planners and by tradespeople, with only passing consideration given to those who must live in the areas being designed and built. All too frequently, these standards are based on the technical and value judgments of the planner, the supplier, the lender, or the producer rather than on the experience and evaluation of the user. The effect has been to impose arbitrarily certain values on user groups as well as patterns of resource allocation geared to professional preference rather than to user preferences. Furthermore, it has been argued that such standards and requirements often exacerbate some of the inequalities in the residential environments available to different population groups. Although these issues have been raised previously, they have not been subjected to research or to rigorous inquiry.

These are some of the relatively unexplored aspects of the neighborhood inquiry that we have sought to rectify. Moreover, the few studies that do exist in these areas are not directed toward public decision-makers and public policy *per se*. We have attempted to fill this gap as well. Accordingly, we have addressed ourselves to many of the questions about the neighborhood unit *de novo*—not just to determine why this paradigm may no longer be valid, but, more important, to piece together some clues that may suggest a new design paradigm.

THE GENESIS OF THE RESEARCH

Candor requires a confession on our part: the scope, orientation, and purpose of this book are slightly at variance with our original research objectives (and those of our colleagues). We have come to the views presented here as a result of engaging in a more full-blown research effort involving a number of our colleagues and students at the University of Southern California. To understand the derivation of our own presentations, it is necessary to sketch briefly the origins and purposes of this larger research effort.

We have already referred to the incorporation of the neighborhood unit concept by the American Public Health Association in the late forties in its publication *Planning the Neighborhood,* as formulated by its subcommittee on the Hygiene of Housing. As one of the first official formulations of ''approved'' residential design, its standards, principles, and assumptions have been incorporated into a variety of subsequent efforts. But it has become an object of criticism because of its very age. Indeed, the organizations most concerned with its use and development—the American Public Health Association and the U.S. Public Health Service—began in the 1960s to express misgivings about its effectiveness. Basically, the criticisms had two aspects: (1) the efficacy of the recommended standards themselves and (2) the use of the ''neighborhood unit'' as the basis for applying the recommended standards.

In view of these criticisms, the Office of Urban Environmental Health Planning of the National Center for Urban and Industrial Health (U.S. Public Health Service) had been concerned for several years about deficiencies in planning standards. Consequently, in collaboration with the Program Area Committee for Housing and Health of the American Public Health Association, it worked on the development of more relevant and meaningful standards for the residential environment. Initially, this effort took the form of seminars, conferences, and position papers presented by authorities in housing, the behavioral sciences, planning and design. The following specific recommendations evolved from these initial discussions.

First, it was apparent that the users of the residential environment—not designers, builders, or financiers—should become the focus of the study, and that user preferences, values, and attitudes should be identified and become an important basis for determining appropriate criteria and standards for residential design. But ''users'' are not homogenous and neither are their preferences. Users include people with different incomes, from different racial/ethnic backgrounds, of different ages, in different stages in their family or life cycle, and so on. Even users with similar characteristics may nevertheless have different values that will shape different preferences. These concerns led to two related premises: (1) The different groups within the population will have their preferences satisfied by differing physical elements (and particularly their attributes, e.g., location and size, *etc.*), of which the residential environment is composed. (2) There are sociopsychological "dimensions" of well-being related to physical and mental health (e.g., safety and privacy, *etc.*) that are partially satisfied through the physical environment, but different groups will satisfy the same dimensions through different physical arrangements or configurations.

Second, an organizing concept such as the neighborhood unit, or any other spatial increment, should not be imposed prematurely on research or on subsequent criteria or standards. As valuable or as intellectually satisfying or intriguing as the neighborhood unit may be to environmental designers and planners, its meaning and utility to the inhabitants of residential environments should be ascertained. If they, too, find it useful, then such an organizing device should be built into the criteria and standards. If not, then some other approach is warranted. In any event, the variety of residential settings and the variety of population groups that exist suggest that any set of criteria and standards should be flexible, so as to accommodate this variety.

Third, flexibility means that design professionals must be able to adapt criteria and standards to differing circumstances. To do so, the professionals must not only be presented with a set of criteria and standards; they must be told how these criteria and standards were derived so as to modify them intelligently. Planners must not depend solely on standards formulated by others, in another place, at another time, for different people with different needs. They must be told how to adapt these standards to the often unique circumstances that each of them faces.

Fourth, the research had to serve two different purposes. First, the findings had to be viewed as illustrative in part—representing the kind of information necessary to formulate better planning standards and criteria—and could not be presented as definitive for all circumstances. The research methodology and its strengths and weaknesses also had to be presented. Such a presentation should be useful to professionals who might design their own research in particular instances to respond to the needs of their own community.

Finally, although user preferences were to be accorded chief concern, the practical consequences of converting these preferences, values, and attitudes into criteria and standards had to be recognized. The standards and criteria had to be useful, understandable, and workable to the producers or suppliers of the residential environment. There are inevitably trade-offs between what the users would like in the abstract and what the producers can build at a price that the users can afford. The standards and criteria based on user preferences had to be formulated with an eye to their costs and consequences to the residential building and financing system as well as with an eye to the public costs of the construction and maintenance of residential environments.

These recommendations and the processes that resulted in their formulation provided the background for the research on the residential environment undertaken by the Graduate Program of Urban and Regional Planning of the University of Southern California in 1969. The first part of this research project was funded by the Bureau of Community Environment Management of the U.S. Public Health Service. The second part, which began in 1973, was funded by the Center for Studies of Metropolitan Problems of the National Institute of Mental Health.

We have summarized the genesis of this research for two reasons: not only to provide a setting and an explanation for the inquiry presented here, but also to provide a circulation of these primary issues that is wider than the confines of our research project. We believe that these issues are important and remain to be investigated, even if we were forced, for financial reasons, to pursue a more circumscribed effort.

In either event, regardless of whether the original research undertaking was fully completed or truncated, there are some basic concerns associated with melding social science with the design arts, and with configuring scientific findings so that they are compatible with the needs of the design professions.

THE ORIENTATION OF THE RESEARCH

The particular form of the research was governed by several factors. As academics trained in the philosophy and methods of social science, we believed that we could improve design practices by making them better informed. Specifically, our approach would be to investigate the behavioral and economic consequences of past design practices so as to better estimate the consequences of recommendations aimed at future efforts.

As planners trained to act in the penumbra of emerging policy issues, where multiple approaches have to be brought to bear on ill-defined problems (cf. Rittel and Webber, 1973), we believed we could put the spotlight of social science techniques on policy areas in residential area design that are particularly in need of research but that are not yet known to pure social scientists.

But as planners trained in design as well, we also knew (sometimes from personal experience) how frustrating it was for designers to incorporate traditional social-science findings into a new design (cf. Baer and Banerjee, 1977). Because the findings of any research inevitably reveal that the phenomena are more complex than was previously imagined, the comparatively simple (but holistic) guides required by designers rarely emerge; in fact, all that does emerge, it seems, are interminable qualifications and provisos—and calls for more research. The facts are so varied and so numerous that the designer, who must synthesize in order to create a solution, is frequently overwhelmed with detail.

Nor has the positivist tradition in social science helped in this regard. The epistemology of design is quite different from the positivist approach, as Schon (1982) has most recently argued. Design is usually based on "personal" or "tacit" knowledge rather than on objective facts (cf. Polanyi, 1958). The compunction in the social sciences to separate fact from value, although diminishing of late, still governs the tenor of "scientific" findings. No alternative that might better assist the design practitioner has yet emerged. Nor has there yet been discovered an external logic that could be applied to sorting this detail for meaningful design. Alexander's *Notes on the Synthesis of Form* (1964) stands as a prime example of how the directly analytical approach to design is ultimately of little help, and Alexander himself later abandoned this approach for a more indirect (but synoptic) one represented by "pattern language" (Alexander *et al.*, 1977).

Thus, we attempted to pursue a research strategy that would be scientifically valid and at the same time professionally useful. This meant, first of all, that in the data-gathering stage we wanted a methodology that would yield results of utility to designers, rather than merely providing material of interest to researchers. Second, in the data analysis stage, we had to consider our findings carefully to determine which ones would prove useful to environmental designers, so as to confine our reporting and comments to those findings. Finally, we could not stop with a display of the data and a discussion of their meaning; we had to go on to present a tentative design paradigm in which the data could be evaluated and used.

We recognize that the residential environment, even though important, is neither the sole nor even one of the most important determinants of the quality of life. For instance, recent evidence suggests that housing and the residential environment, although important, rank considerably below a number of other contributions to a sense of well-being. Marriage and family considerations, work, leisure time, and standard of living are all considered more important (Campbell, Converse, and Rodgers, 1976). Still, as Campbell *et al.* suggested, these more important areas are not particularly amenable to public policy intervention. They defy formulation of successful governmental strategies to improve people's satisfaction in these areas. On the other hand, the residential environment—at least, its physical aspects—lends itself rather readily to government policy formulation and intervention. Furthermore, as Campbell *et al.* (1976) have shown, satisfaction with the residential environment is lower than satisfaction with such domains as marriage and the family, work, and leisure time; a finding that suggests that intervention in the residential environment also holds out the prospect of accomplishing a greater improvement in people's satisfaction with it.

For our purposes, we viewed the residential environment as a sociospatial schema (cf. Lee, 1968); we focused largely on the spatial end of the continuum, and how its planning, design, and management can contribute to the overall quality of life. As much as possible, we attempted to focus on those aspects that are most amenable to control by planners and designers.

THE ORGANIZATION OF THE BOOK

In Chapter 2, we provide a background of the neighborhood unit concept: its origin, its underlying values, the extent of its professional and institutional acceptance, and a review of empirical research that provides some test of its validity. In this review, we focus on some of the key conceptual issues in neighborhood and residential environment research, and thus, we set the backdrop against which our findings must be considered.

Chapter 3 contains a brief description of the people in our interviews, including their demographic and housing characteristics and the residential areas in which they live. Our purpose here is to give the reader a sense of who responded to our questions and the nature of the areas in which they live. The bulk of this chapter is devoted to summarizing the open-ended responses that people gave when asked to describe and evaluate their residential areas. These responses provide perhaps the most encompassing and global sense of what the residential area is all about and some of the key themes that people use in conceptualizing their residential environments. The picture which emerges from these descriptions can be best described as a *milieu* that contains a social, a physical, a functional, and even a symbolic dimension. These open-ended responses also provided a status report for a baseline measure of the current environments, and we have supplemented them with other more objective data drawn from other parts of the interview that we otherwise do not report. In a different way, we have tried to support our interpretation of these open-ended responses by including selected original quotes from the respondents themselves. Finally, we have included some photographs of the interview areas to illustrate these descriptions.

Whereas Chapter 3 reports on verbal descriptions of the residential area, Chapter 4 summarizes graphic representations of individual and collective images of the residential area. By use of the maps drawn by the respondents, we can understand the locational aspects of the *form* of the residential area: the boundaries the residents perceive, the paths they use, the scale and the size of the area they experience, and other factors that cannot be captured adequately through verbal responses. The composite maps, produced from individual maps of each of the twenty-two locations, show some dramatic differences in form characteristics. Furthermore, this mapping approach nicely captures people's perceptions of elements and facilities in the residential area that planners can manage, alter, and organize. In this chapter, we also examine how neighborhood living is valued today and how the concept of neighborhood is related in people's minds to the concept of residential area.

In Chapter 5, we explore still another concept of the residential area—as a *setting* for daily activities. The data presented here come from our respondents' identification of those environmental hardware and activity settings that currently existed in their area, those that were desired, and those that were not. This exploration allowed us to establish a functional list of hardware and settings that together constitute a desirable environment. Here, we also explore the notion of setting aggravation and deprivation that are linked to the overall sense of residential well-being, and how these measures might reflect class inequalities.

Chapter 6 reviews and summarizes the findings presented in the previous three chapters. It integrates the findings, relates them to the issues we pose in this and the next chapter, and discusses implications for public policy.

Finally, in Chapter 7, we develop our arguments for alternative formulations for residential planning and design. In the context of arguments, we briefly review previous attempts by designers and social scientists to formulate alternative schemata and to show why these attempts are inadequate in many ways. We postulate the considerations necessary for an alternative formulation and we propose one, but, more important, we focus on the processes through which such alternatives can be derived. In so doing, we try to develop a normative model of city form that addresses the fundamental issue of inequality and inequity in the distribution of urban resources. In this normative view, we try to integrate environmental design practices with social science research and public policy issues. We conclude by arguing that, like the neighborhood unit paradigm, this normative model represents a confluence of values, but that these values are fundamentally very different from those that made the neighborhood unit so popular at the time of its inception.

In this book we do not specifically consider the formulation of design and the planning standards for future residential environments. However, it should be obvious to the reader that the materials presented in Chapter 3 can easily be considered the basic *performance requirements* of a good residential environment. Materials presented in Chapter 4 provide designers with a number of perspectives for *organizing the future form* of residential areas. Finally, the data presented in Chapter 5 lend themselves easily to use: as a *planning checklist* for appropriate environmental settings and hardware; as *indicators* of residential deprivation and aggravation; and as *location standards* for residential services and facilities. All of these can serve as a basis for zoning and subdivision prescriptions for future residential planning and design.

NOTES

1. For example, see the discussion in Chapter 2.
2. We do not regard innovations in housing or site planning such as the planned-unit-development concept as part of the concern here.
3. By *paradigm*, we mean its established usage (and not in the sense Thomas Kuhn, 1970, used it to describe research in ''normal science'') here, that is, a model, or a pattern that ''functions by permitting the replication of examples any one of which could in principle serve to replace it'' (Kuhn, p. 23).
4. Here, we depart from the company of Alexander *et al.* (1977), who seem to have placed much trust in the ''timeless ways'' of building and design and who appear to have been less concerned about obsolescence and the need for the renewal of ''patterns.'' For a critical review of the ''pattern language,'' see Protzen (1977).

2

The Neighborhood Unit as a Design Paradigm

In this chapter we review the thinking about the neighborhood unit paradigm. Although it is seemingly an oft-told tale among environmental designers, we submit that their story of the neighborhood unit is incomplete, and that frequently they are ignorant of some important intellectual underpinnings of the concept that better explain its configuration. Rather than being a physical design created to accomplish some social ends arrived at *de novo,* the neighborhood unit is actually the three-dimensional expression of some underlying cultural and intellectual beliefs that pervaded American reformist thinking at the turn of the century. Moreover, it is also the most careful summation and delineation extant of more *ad hoc* design practices that have been carried on for thousands of years, for although the precise nature and purpose of the neighborhood unit well illustrates the American penchant for intellectual pragmatism, the roots of the paradigm can be traced back in history to the earliest civilizations. Furthermore, the concept of the neighborhood unit, in addition to nicely capturing some prevalent strains in American thought, has also captured the endorsement of most modern-day societies, for the concept is now employed throughout the world.

The idea of using neighborhoods as a way of structuring, ordering, and presenting the urban society dates to the dawn of civilization. Indeed, Mumford (1954) has argued that neighborhoods are a "fact of nature" and come into existence whenever a group of people share a place. Neighborhood units are as old as the family system and the kinship network in ancient China (Gordon, 1946). The city is symbolized in Egyptian hieroglyphs by a circle circumscribing a formée cross that divides a city into four quarters or "neighborhoods." The tradition of Milesian[1] city planning, in ancient Greece, called for neighborhood units with definite and visible boundaries to serve as primary instruments of social/religious segregation. Thus, the new plan of Thurium (443 B.C.) showed four longitudinal and three transverse arteries that created ten neighborhood units, each assigned to a particular tribe (Mumford, 1961).

The concept of *neighborhood* endured in Roman and medieval cities as well, even though these later neighborhoods were not as meticulously planned physically as their Greek predecessors. A Roman town was typically organized in terms of *vici* (hence our word *vicinity*)—often for administrative purposes—each with its own neighborhood centers and markets. Later, the "quarters" of the medieval towns developed as autonomous units, with their own centers, markets, and sometimes even their own water-supply systems (Mumford, 1961). Although lacking the distinctive boundaries of Greek cities, the medieval city quarters nevertheless had a physical focal point, a sense of place, and, not infrequently, an identity derived from the overlapping organization of parishes.

From this briefest of historical overviews, it is clear that spatial concepts have long been used to organize urban communities. Sometimes, these spatial units have been deliberately created to separate people by family, caste, or ethnic background; sometimes, they have been more the result of other urban activities: the in-between space carved out by major thoroughfares; the imperial convenience of governing by means of administrative areas such as the *vici*; the need to cluster dwellings around a communal water source; and so on. There is also some evidence that "natural" neighborhoods, based on shared occupation, religion, and social networks, were quite common in the cities of antiquity.

Although historical examples are interesting and make the point that people have tended to organize their living and working areas into identifiable units, they are hardly the stuff from which to fashion modern design principles. Scientifically based standards for urban design are not adequately derived from mere precedence and happenstance; historical practice is not a suitable guide for the technologically influenced lifestyles of today; carefully accumulated knowledge of human wants and needs goes unused if professional practice is confined to custom.

Indeed, many of the purposes of these past community configurations are themselves found wanting by present-day values. Ancient cities promoted segregation through a deliberate delimitation of neighborhoods, whereas today the objective is to eliminate impediments to integration. In medieval times, neighborhoods grew organically, without need for approved design concepts, whereas today, consciously guided growth to promote well-being is the principle to which our society subscribes. Accordingly, although we can look to history for examples of past neighborhood development and rudimentary organizing schemes for spatial units, we should seek a more reasoned basis for espousing principles, criteria, and standards for residential design.

THE NEIGHBORHOOD CONCEPT AND ITS HISTORY

Here, we again turn to history—this time to the recent history of the formulation of the concept of the neighborhood unit and its successes and failures. The neighborhood unit approach, first conceived by Perry (1939), was a self-conscious attempt to promulgate good design, and to incorporate the best social thought of the modern era into a physical design that would promote the health, safety, and well-being of people living in urban residential areas. Although it has been widely popular as a design solution, it has also been attacked as being too simplistic and as facilitating social behavior that is now seen as counterproductive to larger national goals.

The basis of the formulation was simple enough: with deliberate planning and design, neighborhoods could be made to be good places to live—and "bad" attributes of urban life could be kept at bay. The task became one of determining those characteristics that would foster good neighborhood development. As we will see, this determination was a strange amalgam of pragmatic philosophy with goals of social betterment and a belief in the means of physical determinism.

Perry conceived of the neighborhood as a geographic unit—a closed system to be used in building-block fashion for the development of urban areas. He proposed that such a unit contain four basic elements: an elementary school, small parks and playgrounds, small stores, and a configuration of buildings and streets that allowed all public facilities to be within safe pedestrian access. To place these four aspects in suitable relationship, six physical attributes of the neighborhood were specified in detail[2] (see Figures 1.1 through 1.3):

> 1. *Size.* A residential unit development should provide for that population for which one elementary school is ordinarily required, its actual area depending upon its population density.
>
> 2. *Boundaries.* The unit should be bounded on all sides by arterial streets, sufficiently wide to facilitate its by-passing, instead of penetration, by through traffic.
>
> 3. *Open Spaces.* A system of small parks and recreation spaces, planned to meet the needs of the particular neighborhood, should be provided.
>
> 4. *Institution Sites.* Sites for the school and other institutions having service spheres coinciding with the limits of the unit should be suitably grouped about a central point, or common.
>
> 5. *Local Shops.* One or more shopping districts, adequate for the population to be served, should be laid out in the circumference of the unit, preferably at traffic junctions and adjacent to similar districts of adjoining neighborhoods.
>
> 6. *Internal Street System.* The unit should be provided with a special street system, each highway being proportioned to its probable traffic load, and the street net as a whole being designed to facilitate circulation within the unit and to discourage its use by through traffic. (Perry, 1939, p. 51)

Although Perry is generally credited as being the author of the concept of neighborhood, there is good reason to believe that he merely crystallized into physical form a number of intellectual and social attitudes prevalent at the time. In effect, he did subconsciously what we are arguing should be done more self-consciously and openly. He took social beliefs and approved cultural mores and converted them into a prototype for planning standards. By exploring the sources of this crystallization, we can gain a better insight into how we might improve our own efforts to rethink some of his basic formulations.

Three broad clusters of values can be identified that have inspired or that have been ascribed to the neighborhood unit concept. The first can be seen as *contextual,* representing the intellectual concerns and thinking that generated the concept. These values are rooted in the social sciences and the humanities, which motivated the reformist thinking in the late nineteenth and early twentieth centuries. The second cluster of values can be called *manifest,* as embodied in the principles of the neighborhood unit concept. These values have traditionally been espoused by practitioners: professional planners, architects, and environmental designers. And finally, the third cluster consists of values that are *tacit* in nature. These are based on pragmatic social and economic considerations, and they are implicit in the endorsement of developers, lending institutions, municipalities, investors, mortgage insurers, and the like.

The Contextual Values

The neighborhood unit formula, although couched primarily in physical-spatial terms, represented an important offshoot of turn-of-the-century reformist thought, which viewed the incipient urban lifestyle of the era with considerable consternation (White and White, 1962). The American intellectual elite was apprehensive about the weakening of the traditional links between the individual, the place, and the community. They were dismayed by the early experience of the evolving industrial city and were uneasy about its inchoate social and moral order. The gradual erosion of the family, the neighborhood, the communities, and the various forms of intimate human associations of the preindustrial era was openly decried. Restoration of interlocking relationships between family, neighborhood, and community was seen as the only way to preserve human dignity, identity, and well-being in the impersonal environment of the urban agglomeration. These sentiments were intellectual antecedents to the development of the neighborhood unit concept.

Proponents of these views, however, were not necessarily resigned to seeing a total bankruptcy in the moral and social order of the city. Many were pragmatically optimistic about the developing metropolis; this optimism has been embodied, for instance, in the view of psychologist and philosopher William James. Indeed, it was James's views that inspired Jane Addams (a social worker), Robert Park (a sociologist), and John Dewey (a philosopher and educator), to try in their own separate ways to preserve these humanistic ideals in an urbanizing society (White and White, 1962). The efforts of each of these people were to become important cornerstones of the subsequent development of the neighborhood concept.

The basic tenet of the reformist movement was not to defy or to ignore the city but to instill a spirit of neighborliness, community, and social communication in its seemingly hostile and anomic environment. The settlement house[3] and the school were seen as institutions that could nurture the future growth of community and neighborly feelings (Dahir, 1947; Gallion and Eisner, 1975). Thus, Jane Addams, herself a founder of a settlement house, emphasized it as a means of establishing communication and community among oppressed and apathetic city-dwellers. John Dewey prescribed the community-based school of the preurban vintage as a model for urban education. Robert Park advocated a "new parochialism" in the form of local communities to establish the identities of the primary groups and to protect the individual from the loneliness of city life.

In retrospect, it is apparent that the emerging city of the late nineteenth and early twentieth centuries introduced changes into the prevalent lifestyle that were unprecedented in urban existence and that were perceived as too threatening in impact. By attempting to preserve preurban values and social institutions in an urbanizing context, the reformist movement was searching for ways to protect the individual from the current waves of rapid change characterizing that period. The revival of the neighborhood and the local community was essentially an adaptive mechanism to cope with a phenomenon that Alvin Toffler (1970) was to describe many years later as "future shock."

These social values were closely paralleled by another reformist view that promoted the cause of the local neighborhood by emphasizing its political potential in a grass roots democracy. According to Guttenberg (1978), the leading proponent of this view was a Presbyterian minister named Edward Ward who advocated the use of neighborhood schools as adult social and civic centers. Aside from their civic and cultural purposes, schools were seen as the seat of neighborhood democracy, representing a form of neighborhood government; symbolically, they became "neighborhood capitals," standing as it were in opposition to the tyranny of city hall. Small wonder, then, that schools were later to become the physical centers around which neighborhood units were geographically (and symbolically) formed.

More significantly, it was the theoretical arguments of the social-psychologist Thomas Horton Cooley that further strengthened these converging ideals. He argued that families and neighborhoods are the main "nurseries" for healthy personality and social development (Guttenberg, 1978). Cooley's emphasis on the importance of primary group association to the social and moral survival of the urban man was to have a profound impact on the protagonists of the neighborhood unit principle and would be frequently invoked in its defense (Dahir, 1947).

The Manifest Values

These moral, social, and political values converged on the concept of the neighborhood and contained the seeds for its physical manifestation in the neighborhood unit concept. It is clear from Guttenberg's (1978) recent account of the evolution of Perry's design ideas that the widespread use in the evening of public schools for recreational, social, and cultural activities played an important role in giving the elementary school pivotal importance in the physical concept of the neighborhood. And it was the school that helped to establish other important parameters, such as the walking-distance radius of the neighborhood and the population size, based on the optimal service population needed to support a school. The need for boundary definition and for the integration of recreational areas in a centripetal spatial organization also grew out of the same basic social and political views.

Perry saw the neighborhood unit as a residential cell in the "cellular city," which he argued was an inevitable product of the automobile age. Thus, like Park (1952), Perry identified the automobile as a chief villain in the emerging way of life. Whereas Park and his colleagues made a plea for primary group communities to establish a social and moral order, Perry envisaged neighborhood units as the basis for introducing physical organization and order into the urban form. Indeed, in an age that nourished a belief in physical determinism, it was a logical step in translating ideas into action—in using architectural design to accomplish social ends.

The cause of the neighborhood unit was further enhanced by certain complementary paradigms already being promoted at that time by a group of reformist planners and designers. Inspired by the writings of Patrick Geddes and Ebenezer Howard and influenced by the works of Barry Parker, Raymond Unwin, and Frederick Olmstead, these planners and designers had sought to develop new forms of residential area planning and design that included concepts of "garden cities" and "superblocks" (see Figure 1.2 and Figure 1.3). Indeed, as Mumford (1951) has pointed out, even before Perry's neighborhood ideas were fully crystallized, Clarence Stein and Henry Wright had already included many of the principles of the neighborhood unit in their design of Sunnyside Gardens in New York. Nevertheless, once Perry's formulations were firmly established, the idea of neighborhood unit nicely dovetailed with the other two design concepts, both in spirit and in principle. This "troika" of design paradigms provided the conceptual foundation for the subsequent new community developments, including the famous Radburn Plan. Indeed, Radburn may have been the very first case where the neighborhood unit formula was applied and realized in the form that is now widely accepted (Stein, 1957).

The design values implicit in promoting neighborhood units in those newly developing communities were very much rooted in the same humanistic ideals as those advocated by Addams, Dewey, and Park. These values responded to the same sentiments by attempting to create a physical place that was consistent with the sense of a community: a setting that provided opportunities for leisure, recreation, and social interaction, and an environment that was safe, protected, pleasing, and secure. It was believed that through the physical design of the settings, it would be possible to reinstill in the individual the sense of dignity, freedom, and identity that were being threatened by the massive urban agglomerations. To be sure, in their formal statements, the designers emphasized comeliness, convenience, spatial layout, recreation spaces, and separation of pedestrians from traffic; but the above social and moral objectives were the driving force for justifying these design principles.

The Tacit Values

It is interesting to note that the physical design of the neighborhood unit had always presumed a social homogeneity by prescribing uniformity of dwelling types. It was clear to many supporters of the neighborhood unit that it was meant for families with children (Wehrly, 1948). Childless couples or single individuals were not expected to seek neighborhood living. But the underlying presumption of social homogeneity was gradually expanded beyond stages in family cycle to mean ethnic and income homogeneity, too. At least, this is what critics like Isaacs (1948a) and Bauer (1945) claimed, pointing to the restrictive covenants recommended by the Federal Housing Administration's underwriting manuals. These covenants were recommended to protect social and income homogeneity on the assumption that incompatible groups in a neighborhood could lessen or destroy owner-occupancy appeal (Federal Housing Administration, 1947). Thus, social homogeneity was seen as the best guarantee for protecting the interests of the producers and lenders.[4] Not only did the concept of the neighborhood unit assure this social homogeneity, but its design also guaranteed the territorial integrity and impermeability of a particular social group. Indeed, the neighborhood unit concept won approval not just from the FHA but from other government agencies as well. And, of course, it is no surprise that it won acclaim from professional real-estate operators, leaders of chambers of commerce, and other business groups (Dahir, 1947).

Thus, the values underlying the neighborhood unit concept have been diverse. The values of family, neighborliness, community, and group identity have been paramount, but ideals of grass roots democracy and community control have also played roles. Additionally, goals of personality and social development, as well as civic responsibility have been a part of these contextual values. Safety, security, appearance, beauty and visual identity, naturalness of surroundings, and overall physical-spatial order are some of the components of the manifest values.

But the popularity of the concept and its widespread use in practice (which we describe in the next section) cannot be attributed solely to society's embracing of these humanistic values. Support of these values through the use of the neighborhood concept promoted some economic values as well (Churchill, 1945), which thereby contributed to its endurance as a widespread concept. Thus, preservation of the home owner's property values, protection of the mortgage lender's security, and maintenance of the community tax base were the pragmatic values of the marketplace—the ultimate supplier of residential areas—which the neighborhood concept also served in the course of meeting humanistic values. But whereas these economic (market) values were nicely served, concern about social equity and the fair allocation of public (and private resources) has never been prevalent.

THE USE OF THE NEIGHBORHOOD CONCEPT IN PROFESSIONAL PRACTICE

Although there is a rich literature on the pros and cons of the neighborhood concept, there has been little investigation into its use. One would think that, in view of such strenuous debate, the principle of neighborhood units would be used only sparingly. Quite the opposite is true. The neighborhood concept is ubiquitous, both in the United States and abroad—it has no rivals or even challengers. The formulation is so powerful (despite its shortcomings), and apparently design professions are so in need of a model (despite their claims to creativity), that the neighborhood concept has been adopted and adapted throughout the world.[5]

Twenty years after Perry had formulated the neighborhood unit concept, the prestigious American Public Health Association adopted it in its publication of planning standards, *Planning the Neighborhood,* which has had widespread use. In publishing this document of approved professional standards, the American Public Health Association stated several reservations about the validity of the standards. Because these standards were based more on the opinions of experts than on scientific research, they were replete with normative judgments; and because the standards had been quantified in areas that could not be precisely measured, the precision of knowledge implied was artificial.

Despite these caveats, the document was instantly influential for three reasons: first, it had been developed by experts; second, it presumed to link the built environment with concerns of health and hygiene; and third, and perhaps more important, because no other "authoritative" standards were available. The document became widely used, and in legitimizing Perry's neighborhood unit concept, it influenced a variety of professional guides.

In laying out residential developments, architects, engineers, planners, and landscape architects all refer to books of standards published by their professional organizations. Housing developers use guides issued by the Urban Land Institute and published in trade journals such as *House and Home* and the *Journal of the National Association of Homebuilders*. Lenders use guidelines published by their associations, such as the U.S. Savings and Loan League. Insurers of loans use regulations and manuals published by the U.S. Department of Housing and Urban Development, the Federal Housing Administration, and the Veterans Administration. Apparently, all have been influenced by the neighborhood concept.

In examining the published documents of twenty-three such organizations, Solow, Ham, and Donnelly (1969) determined that, where evidence could be found, each organization supported all or at least some of the principles of the neighborhood unit concept. None indicated opposition to the concept (see Table 2.1). As these authors stated,

> Such influential associations as the Urban Land Institute, the American Society of Civil Engineers, the American Institute of Architects, the American Society of Planning Officials, the International City Managers Association, and such worldwide agencies as the United Nations clearly advocate, though not dogmatically or inflexibly, the neighborhood concept, very much along Perry's principles.
>
> Another important group such as the National Association of Home Builders, the AIP, N[A]HRO, IFHP as well as Federal Agencies, HHFA (now HUD) and FHA advocate, in general, some form of neighborhood unit, and some principles of components of Perry's concept, although the majority makes [*sic*] no direct reference to Perry. A number of these focus on desirable physical attributes and standards of neighborhood design without advocating Perry's unitary concept. (p. A-1-25)

Practitioners confirm the high utilization rate of neighborhood standards as suggested by their ubiquity in professional publications. In a survey of 258 members of the American Institute of Planners (now the American Planning Association), Solow *et al.* found that half the group thought the neighborhood unit concept useful, valid, and ideal for public policy. Nearly 80% used the concept in practice. Indeed, more than 55% used the specific document *Planning the Neighborhood* (PTN) in their practice. However, some members expressed reservations. For example, 15% did not believe that the concept was useful or valid; 33% found it less than ideal; and approximately 20% did not agree with the concepts of the standards embodied in PTN (Solow *et al.*, 1969).

Exploring other dimensions of the use of the concept, Solow *et al.* once again found evidence of its ubiquity. Although the concept was not a *legal*

TABLE 2.1. Organizational Advocacy of the Neighborhood Unit Concept[a]

Name of organization	Level of support: Clear support	General advocacy	Lack of available statement	Clear opposition
American Institute of Architects	X			
American Institute of Planners		X		
American Society of Civil Engineers	X			
American Society of Planning Officials	X			
Central Mortgage and Housing Corp.–Canada		X		
Department of Housing and Urban Development		X		
Federal Housing Administration		X		
Housing and Home Finance Agency		X		
International City Managers Association	X			
International Congress for Housing and Town Planning		X		
International Federation for Housing and Planning			X	
National Association of House Builders		X		
National Association of Housing Officials		X		
National Association of Real Estate Boards		X		
National Federation of Settlements and Neighborhood Cities		X		
National Housing Agency–Federal Public Housing Authority	X			
National Housing Center			X	
National Housing Conference			X	
Town and Country Planning Assoc.	X			
Town Planning Institute			X	
United Nations	X			
U.S. Chamber of Commerce			X	
Urban Land Institute	X			
Total	8	10	5	0

[a]Source: Solow, Ham, and Donnelly (1969), Table 1, p. A-1-26. Used with permission.

requirement in subdivision regulations, there was evidence of its endorsement by local government handbooks issued to assist designers and builders in planning residential subdivisions. Moreover, it was frequently embodied in city general plans and endorsed in urban renewal programs and guidelines. Finally, although local zoning ordinances do not generally lend themselves to encompassing the neighborhood unit concept, the application of the newer Planned Unit Development in lieu of traditional zoning requirements does frequently permit its use.

In the United States, the neighborhood unit concept has been used in new town developments from Radburn to Columbia, in mass subdivisions like Levittown, and in more piecemeal form throughout the nation. The concept has also been used in new towns in England, Israel, Sweden, and the Soviet Union (Keller, 1968; Porteous, 1977).

Since the Solow, Ham, and Donnelly study, the neighborhood unit has reappeared in another form, attesting once again to its inherent viability. In 1972, a National Policy Task Force of the American Institute of Architects (AIA) advocated use of the "growth unit" in the building and the rebuilding of the nation's cities. Whereas the neighborhood unit was largely formulated from the demand side (that is, from the perspective of the user of the residential area), the growth unit was formulated from the supply side (from the perspective of architects and developers). Despite these differences, there are marked similarities between this growth unit and the neighborhood unit concept:

> The growth unit does not have fixed dimensions. Its size in residential terms normally would range from 500 to 3,000 units—enough in any case to require an elementary school, day care, community center, convenience shopping, open space, and recreation. Enough, too, to aggregate a market for housing that will encourage the use of new technology and building systems. Also enough to stimulate innovations in building maintenance, health care, cable TV, data processing, security systems, and new methods of waste collection and disposal. Large enough, finally, to realize the economies of unified planning, land purchase and preparation, and the coordinated design of public spaces, facilities, and transportation. (AIA National Policy Task Force, 1972, p. 4)

And like the neighborhood concept, the growth unit can be designed and evaluated as a unit or package instead of as a disjointed collection of activities and undertakings in the residential environment.

In summary, the neighborhood concept and the document *Planning the Neighborhood* have been widely used by practitioners and are evident in a variety of professional publications. They have been extremely influential in structuring professional thinking and in determining the content of subdivision regulations and zoning ordinances. The pervasiveness of their influence and use, despite professional reservations and intellectual repudiations, suggest the power of neighborhood unit as a construct.

SHORTCOMINGS

Initial criticisms by social scientists were directed at the practitioners' preoccupation with the physical aspects of the neighborhood unit: access, layout, boundaries, segregated land use, overall appearance, and so on (Dewey, 1961; Tannenbaum, 1948). Implied in this criticism was the argument that the social neighborhood and neighborliness may not necessarily result from the spatial layout of the physical neighborhood; social homogeneity may be more important (Mann, 1958; Keller, 1968; Lefebvre, 1973; Pahl, 1970). Although the critics were troubled by the physically deterministic tenet of the neighborhood unit concept, they did not dismiss it outright. Rather, many of these critics felt compelled to offer social justification for the concept while warning planners against its uncritical application (Tannenbaum, 1948; Dewey, 1961).

The severest of the criticisms of the concept, interestingly, came from the planners themselves. Reginald Isaacs (1948b, 1949), for instance, challenged the validity of the neighborhood unit in the urban context, charging that it was ostensibly inspired by nostalgia for a rural lifestyle. He further asserted, along with other planners (Bauer, 1945), that even if the concept were valid, it was tantamount to an instrument of social and economic segregation.

In more recent years, the neighborhood unit concept has also come under attack on the philosophical level as being anachronistic, because in the emerging lifestyles of the postindustrial society, "nonplace" urban realms are taking precedence over primary group associations and "place" communities (Webber, 1963, 1964; Webber and Webber, 1967). From a more pragmatic standpoint, it has been argued that the physical scale of the neighborhood unit as originally conceived is not adaptable to the present increments of residential development and redevelopment.

The neighborhood unit and the related planning standards were subject to further criticisms stemming from the "cultural relativism" argument. Both the intellectual elite and minority leadership have asserted that the concept of a "great American melting pot" is essentially a myth (Glaser and Moynihan, 1963), and that different ethnic values, lifestyles, and preferences must be acknowledged and preserved (Berger, 1966). The neighborhood unit concept and the related standards were challenged as being a product of white, middle-class professional planners—and thus a symbol of white, middle-class values. Whether something other than neighborhood living was preferred by the other income or minority groups was not discussed, but there was an implication that the residential environments for different social groups would follow models that were based on their different interests, hopes, aspirations, and desires.

But these early criticisms—like the neighborhood unit concept itself—were largely speculative and conjectural. There was no empirical test of the concept to determine if it fulfilled the ascribed purposes. Because the concept was widely embraced by practitioners, it is not surprising to find that a number of studies have investigated its success in use. Although any number of residential developments have elements of the neighborhood unit concept embedded in their design, the purest examples are typically found in new town developments both in the United States and abroad.

The record of the neighborhood unit in the United States is mixed. A comparative study of planned and less planned residential communities by Lansing, Marans, and Zehner (1970) included Radburn, Columbia, and Reston, three well-known American new towns where the neighborhood unit concept was utilized in organizing and planning residential areas. According to this study, the planned features ("the general community plan or idea," which presumably subsumed the neighborhood unit concept) of the new towns was an appealing factor in decisions to move for 51% of the respondents in Columbia, 36% in Reston, and 18% in Radburn. All three communities received good to excellent overall ratings from most respondents, in particular from families with young children and from the elderly. Teenagers, however, gave poor ratings to those communities, reflecting what the British sociologists refer to as the "new town blues." Indeed, according to their evaluation, the less planned communities received better ratings than the planned communities. Although some of the physical amenities for children in the planned communities made them appealing to families with children, these "planned" features were mentioned much less frequently as a source of satisfaction than were accessibility to jobs and quality of schools and neighbors. In these communities, the neighborhood unit as an element of planned development played a minor role in community satisfaction.

Similarly, a more recent study of ten new towns in the United States (Burby and Weiss, 1976) found that, although the neighborhood unit concept had influenced the residential area design of these communities, these planned features played a minor role in the residents' satisfaction.

In a study designed to find out why people purchase residences in planned communities, Werthman, Mandell, and Dienstfrey (1965) found that some of the features of the neighborhood unit concept were not actually liked by the residents. For example, the planners' insistence that shopping and community facilities be the focal point of the neighborhood conflicted with the residents' perceptions of undesirable effects associated with such land uses. In fact, most residents preferred to have such facilities, including schools, located on the periphery of their residential area. The study also found that the respondents were quite skeptical of the social goal of interaction among the residents as idealized in the neighborhood concept. Furthermore, "planned" characteristics were of concern in purchase decisions only to the upper middle class. The authors pointed out, however, that it was not that the lower-income groups did not appreciate the aesthetics of planned development (in fact, they complained about the monotony of their boxlike houses), but that they usually associated such an aesthetic environment with more expensive housing that was beyond their reach.

The neighborhood unit concept failed to get even minimal support in Gans's 1967 study of the "ways of life and politics" in Levittown. Gans concluded that the traditional neighborhood scheme that influenced the design of the residential areas did not affect people's lives or social relations. He attributed this failure to the size of the neighborhoods (too large for social interaction, as Willmott, 1962, had earlier noted in the case of Stevenage, England); their lack of distinction (too similar to each other to develop a sense of identity); and an absence of the social and political functions that usually help to coalesce neighborhood spirit. Even the elementary school, located in the middle of the neighborhood in Levittown, failed to serve as a focal point.

In the case of the British new towns, the data suggest that the neighborhood concept may have helped the planners to conceptualize a hierarchy of residential subunits and to determine the size and the location of various community facilities, but that it had little congruence with the residents' existing social structures (Willis, 1969). Willmott (1962) concluded from his survey of Stevenage residents that, although the neighborhoods seemed to work in functional terms, there was little evidence that they created a sense of "community" or "neighborliness," or that they had any special social significance. He found that the physical boundaries of the neighborhood had little to do with socialization patterns; intense socialization usually took place within a much smaller area.

A separate survey of the residents' travel patterns in Stevenage essentially echoed this conclusion. It showed that, although the neighborhood structure accurately reflected school and weekday shopping trips, it did not reflect the patterns of social and recreational interaction (Bunker, 1967; see also Salley, 1972).

In a similar vein, Garvey (1969) concluded from resident activity patterns that neighborhood concepts did not function as ideal settings for social interaction. And from a comparative review of ten neighborhoods of British new towns designed after neighborhood unit principles (but with some modifications), Goss (1961) observed that a wholly satisfactory interpretation had not yet been reached in terms of optimal size or mix of housing and service areas.

This is not to say that the residents of the British new towns could not identify with a spatial neighborhood. Willmott's (1967) later study of Ipswich showed that most of its residents were able to identify some of the boundaries of their "neighborhoods." Although there was great variation in the sizes and boundaries of such neighborhoods, there was some consensus on major boundaries. In fact, even in Cumbernauld, where the neighborhood concept was deliberately abandoned, there was some evidence that the residents were able to develop a sense of identity with different residential subareas (Godschalk, 1967). The point is, as Willmott (1967) argued, that, although most people can identify some sort of neighborhood, their congitive maps bear little resemblance to the conventional neighborhood unit model.

Questions have recently been raised about the functional adequacy and efficiency of the neighborhood unit concept. Based on his observations in Columbia, Slidell (1972), suggested that the neighborhood concept is financially inefficient. The currently prescribed size and density of the neighborhood unit does not always provide the critical mass necessary to support some of the local services and facilities. Furthermore, the economic viability of such local services and facilities is often undermined by a duplication of services at the town center level, as is shown by Low (1975) in the case of Cumbernauld.

The neighborhood unit has also been criticized as an inflexible planning scheme. As Herbert (1963) argued, the neighborhood unit "stands for commitment—commitment to a fixed school policy, a fixed shopping system, an unchanging pattern of use" (p. 190). The failure of the neighborhood schools to meet larger community objectives is becoming apparent as communitywide busing is required by federal court order in some communities, even though the neighborhood school has become a symbolic rallying point for the antibusing groups. At least one school district, the Wake County School Board in North Carolina, has indicated that neighborhood identification would be deliberately discouraged in locating future sites (see Slidell, 1972). In Boston, a serious evaluation of school facilities was undertaken in light of changing demographic profile, neighborhood configurations, and busing needs.[6] Thus, a planning concept that is dominated by a doctrine of neighborhood schools has clearly become out of phase with changing social situations.

The empirical studies of planned neighborhoods raise serious questions about the congruence of physical area with social networks, and about the degree of satisfaction that can be derived from physical design. These questions had been raised earlier by sociologists who, intrigued by this initial discourse by planners and following up on some earlier work by Park (1952), investigated the basic notion of the neighborhood to determine its empirical validity. Rather than assuming (like the early planners) that neighborhoods exist, and that investigation should be aimed at determining their best physical form, these social scientists were interested in the more basic question: Do neighborhoods exist, and if so, how do we recognize them? And further: Even if they do exist and we can identify them, are they necessary for the overall satisfaction and well-being of urban dwellers?

Clearly, these kinds of questions should be answered before it can be determined whether the neighborhood unit concept should be revised and, if so, how. So far, however, the findings of social scientists have not resolved these questions, although they have raised the level of understanding of and sophistication about the topic. A host of differing interpretations and perspectives are offered today on the definition of *neighborhood,* even though the term is so commonplace as to cause seemingly little misunderstanding. Because of its etymological link with the word *neighbor,* it has always carried with it the social perspective of relationships between people. But in urban contexts, the geographic orientation of propinquity and territoriality is increasingly stressed. As modern researchers have attempted to make the term *neighborhood* more precise and more amenable to description and measurement, they have been confronted with the variety of ways in which the two basic concerns—people and territory—

can be combined. Sociologists, for example, have examined the differences between being friends and being neighbors and have looked at the interaction patterns of neighboring within urban communities. Environmental designers, on the other hand, have limited their observations to how people use space and facilities within a geographic area (Keller, 1968). Thus, neighborhoods have been defined in several ways: by the characteristics of people residing in proximity; by boundaries (social, physical, symbolic, or demographic) that people use to distinguish between areas; by the activities within an area; by the function an area serves; or by some combination of these. All of these approaches have shed light on the concept and have simultaneously defied efforts to come to an agreement on a single definition. The lack of a single definition also makes it difficult to test the appropriateness of the neighborhood unit concept.

CONCLUSIONS

What have we learned from this review? The skeptic might conclude that the major message is "more is less": the more we investigate this topic, the less practitioners know what to do; the more sophisticated we become about the subtleties and the various dimensions of the notion of the neighborhood, the less able we are to provide clear-cut guidelines or principles of design to assist the planner.

There is an element of truth in this conclusion, but the mistake lies in expecting the wrong kind of answer to emerge from the inquiry. Rather than expecting social science research to seek design solutions to problems that have no definitive answers, we must look for design proposals or models that are not contrary to what the research has revealed, so that tentative solutions—when proffered, although not optimal—at least will not result in grievous mistakes. This is essentially a strategy of minimizing misfits rather than finding a best fit (Alexander, 1964).

This review reveals that, although the neighborhood design concept was innovative and captured some beliefs that were prevalent in the United States at the turn of the century, it has increasingly lost its potency as a present-day design concept. It may still have theoretical validity and could be applied to situations where the scale of development and the aspirations and lifestyles of the consumer groups make it suitable. But as Solow *et al.* (1969) noted, it is neither necessary nor desirable to regard it as the only model or as a universally appropriate one. Rather, it should be regarded as one of many other types of possible residential environmental units. But as we will argue later, based on our findings presented in the next three chapters, it is difficult to support the model even as an alternative in the face of issues of equity.

It is therefore incumbent on the community of environmental researchers and designers alike to provide new design paradigms to respond to the objections to the old one and to stimulate further creativity by providing new alternatives. One way to provide these alternatives is to break away from the very concept of neighborhood itself. It will be recalled in our earlier review of the literature that many sociologists have maintained that the neighborhood is more a sociological than a spatial phenomenon. It can be further argued that if we uncouple the physical model from the ''neighborhood'' concept (which is basically sociological), we avoid the problem of having to accept or reject the original concept and any of its intellectual baggage that seemingly impedes future innovations. Instead, we can study the problem afresh.

We begin the search for a new paradigm by adopting a new focus: the *residential environment.* It can be defined, after Solow *et al.* (1969), as

> The land, facilities, services, and social structure which supplement the home in providing for satisfaction of individual and family needs, social interaction, personal development, and political participation and which delimit the territory appropriately included in the design of a residential environment. (p. A-1-47)

Such an umbrella definition can capture the diversity of locations, environments, and consumption patterns of a pluralistic society without being fixed to any inviolate model. Unencumbered by the meanings (and increasingly ''loaded'' connotations) of the term *neighborhood,* this reorientation provides a fresh conceptual start and also offers a more relevant frame of reference for research. In the next three chapters, we discuss the context and the selected findings of our study of user perceptions and preferences related to residential environments in the greater Los Angeles region. These findings, in turn, have helped to suggest new grounds for a reformulation of the physical model for residential planning and design, which we present in the concluding chapter.

NOTES

1. After the tradition represented by the plan of the ancient Greek city Miletus.
2. It should be noted that, in this original formulation, shopping was not included in the heart of the neighborhood unit. Later, as the neighborhood unit concept began to be identified with the service area concept, neighborhood shopping was shown in the core, especially in some of the new town development concepts. The neighborhood unit model proposed by Clarence Stein showed this feature (see deChiara and Koppelman, 1975, p. 265).
3. Settlement houses, as precursors of contemporary models of senior citizens' centers, social service centers, missions, retirement homes, and so on, served similar welfare functions in the congested poor areas of the city, but with a much broader range of clientele and services. Typically, settlement houses provided various educational and recreational services for the entire community.
4. Dahir (1947) quoted the following statement from a Federal Housing Administration Bulletin

3

The Research Instrument and Respondent Impressions of the Residential Environment

The neighborhood unit, its intended purposes and subsequent criticisms, were the starting place for our own efforts. As mentioned at the outset, however, we did not investigate the neighborhood unit *per se* based on our assessment of its strengths and weaknesses. Rather, we were able to examine its derivatives as manifested in existing built environments. Accordingly, we substituted the "residential area" as the subject of most of our queries to our respondents, hoping to avoid any loaded connotations or biases stemming from the use of the word *neighborhood*. Moreover, instead of telling our respondents what the term *residential area* encompassed, we let them tell us what it meant to them, so that we did not prejudge or predetermine the outcome. In this way, we hoped to tap the essence of the residential experience.

The various stages and aspects of the research that we report in the following chapters are summarized in Figure 3.1. Here, we show the initial means of focusing the research, the derived residential formulation, the survey instrument based on this formulation, the variety of population groups that we interviewed in twenty-two residential areas in the Los Angeles area, the data analysis (and a comparison of the "neighborhood" with the "residential area" as perceived by our respondents), and the final formulation of an alternative paradigm based on our data analysis and literature review.

THE SCOPE OF THE RESEARCH

As the orientation of our undertaking was manifold in an area that had historically been subjected to narrow research, we used a research instrument

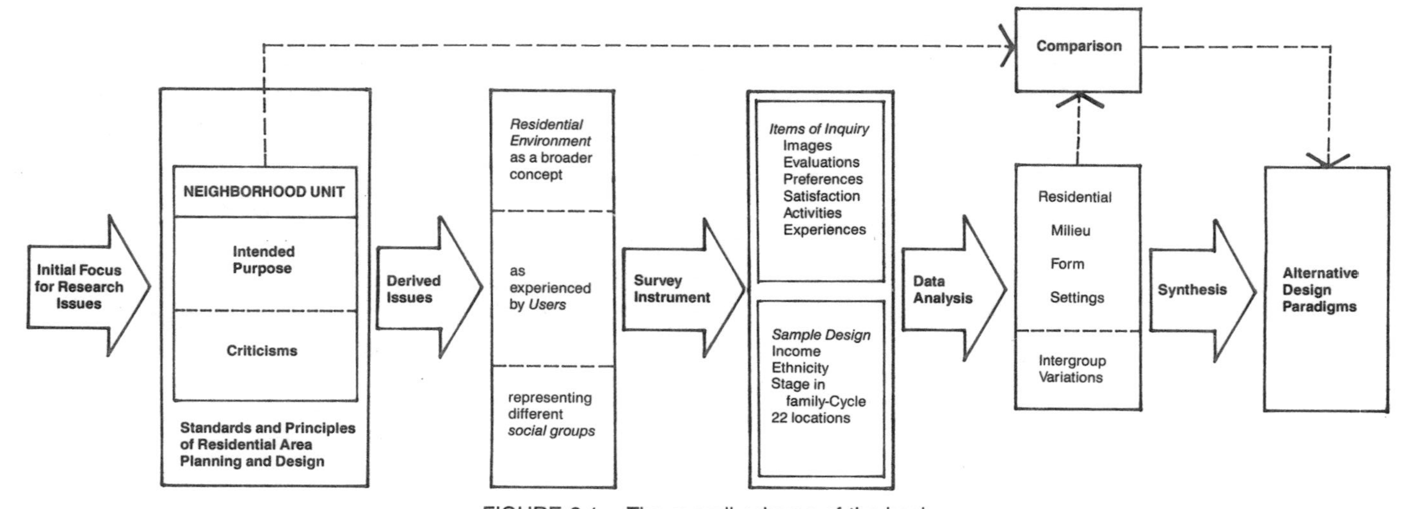

FIGURE 3.1. The overall scheme of the book.

composed of numerous parts and approaches. Some of these approaches were tried and true, having been used extensively by researchers in the past, although not necessarily for the purposes to which we put them. Others were original with us—our attempts to provide data in areas previously unexplored. In all of this, we sought not only the data to be derived from the methodologies, but also an evaluation of the effectiveness and the appropriateness of the methodologies themselves. Our findings about population groups in the Los Angeles area may not apply strictly to every other part of the nation, and hence, we hoped to provide guides to methodologies that could be used elsewhere.

In its totality, the research instrument consisted of six basic parts (see the interview schedule, Appendix I): a verbal description and evaluation of the residential area; a pictorial depiction of the area with questions about the maps that the respondents drew; a semantic differential procedure; a listing and evaluation of environmental settings; a trade-off game for choosing a more desirable combination of environmental attributes and their preferred levels; and a description and evaluation of residentially related activities.

We have grouped them into two sections (Figure 3.2) to distinguish better the findings we present here from the totality of the findings collected. As mentioned, design paradigms require some composite whole if they are to be of use to designers; bits and pieces that are not well connected to anything else are of little use to them. Thus, the first section of three methodologies (on the left in the figure) consists of those that we report here because (1) they produced findings that could be used directly by designers, as they were of a whole; and (2) they could be appropriately used by localities that wish to conduct their own research to devise residential standards for their community. These methodologies attempted to elicit what the residential area means to its inhabitants, what it consists of, and how it is evaluated.

The second set of methodologies (shown on the right in the figure) produced findings that were more indeterminate and incomplete. These methodologies dealt with how residential areas were perceived and used and with how they could be improved. They were initially devised because they were thought to more exactly determine specific standards that could be used in the design of a residential area. We did not think the findings from these methodologies in their present state would be suitable to inform design. The results derived suggested methodological problems that remain to be worked out. These findings are nevertheless of interest to social scientists, and have been reported separately as progress reports (e.g., Banerjee and Flachsbart, 1975; Banerjee, Baer, and Robinson, 1974; Flachsbart and Phillips, 1981; Robinson, Baer, Banerjee, and Flachsbart, 1975).

It will be recalled that a major criticism of the neighborhood unit formula was that it failed to consider the variable values and preferences of different consumer groups. This argument is rooted in the premise of cultural pluralism,

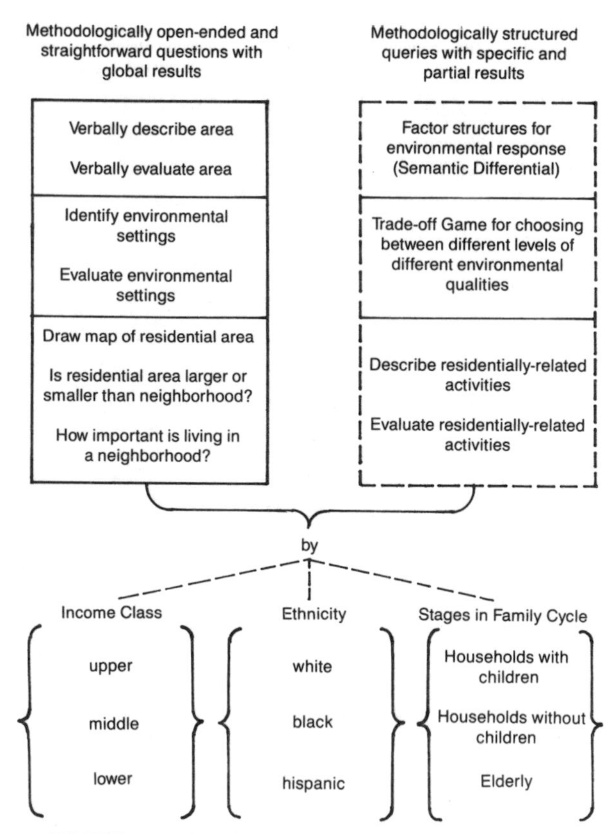

FIGURE 3.2. Scope of the research and methodologies.

which requires that the values, customs, needs, and preferences of various segments of society be respected and accommodated in public policy (Berger, 1966). In this context, it was proposed that residential settings should be designed to reflect the heterogeneity of the consumer groups rather than the "typical" or "average" characteristics of the populace as a whole.

THE RESEARCH DESIGN

Our own research, therefore, was aimed at responding to those kinds of criticisms. To explore whether plural norms and values should be used, our sampling plan stratified respondents along three dimensions—our major independent variables:

1. *Race/ethnicity*. The categories chosen were white, black, and Hispanic (Spanish surname).[1]

2. *Income*. Household income was divided into three categories: upper, middle, and lower—designations commonly used by practitioners. The exact categories (in 1970 dollars) are shown in Table 1, Appendix II.
3. *Stage in the family cycle*. Three categories were used: (a) families with children; (b) families without children (living with them) and with the head of the family under 62 years—the "childless"; and (c) families without children (living with them) where the head of the family was over 62—the "elderly."[2]

Figure 3.3 illustrates the sampling scheme.[3] Upper-income blacks and Hispanics were not included in the survey because they are such a disproportionately small percentage of the upper-income groups that they are exceedingly hard to locate in efficacious sampling clusters. Lower-income blacks and Hispanics are easier to locate in clusters but are more reluctant to grant interviews. We translated our interview schedule into Spanish and used Spanish-speaking interviewers for the lower-income Hispanic areas. We used special interviewers familiar with the lower-income black areas so as to obtain interviews with that group of respondents.[4] At times we had to depart from the strict sampling protocol used with the other five groups in order to obtain interviews at all. Therefore, the data from these two groups must be treated with some caution.

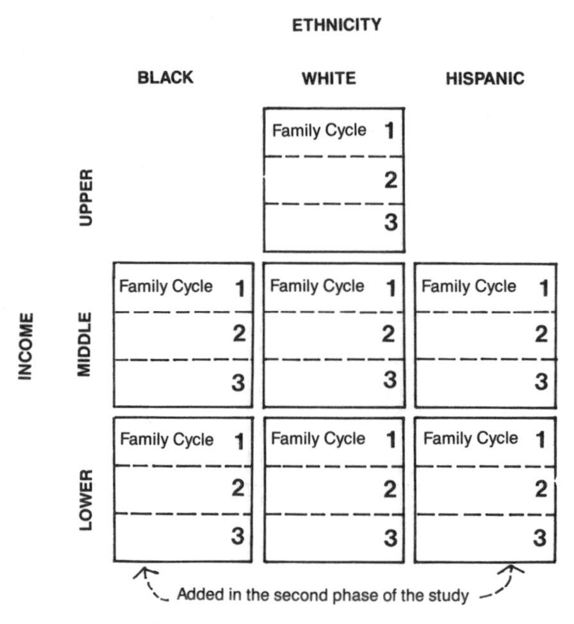

FIGURE 3.3. Sampling scheme.

For each of the seven population groups shown in Figure 3.3, we selected two to four different, widely separated geographic areas in the Los Angeles metropolitan area from which to sample for respondents.[5] The twenty-two locations from which we derived 475 complete interview schedules are shown in Figure 3.4. This approach allowed us to mitigate the influence of unique physical characteristics within any one interview area. It also allowed us to collect independently some "objective" data on the areas that the respondents would describe and evaluate.

Within these areas, households were picked at random[6]; a letter was sent to them explaining the project; and they were then interviewed by persons trained to establish a quick rapport with these potential respondents while determining if they were "eligible" (i.e., met our three criteria of race, income, and stage in the life cycle for that area). If the household members were eligible and willing, the member of the household to be interviewed (also chosen at random from household members sixteen years old or older) was asked for an appointment for a 2.5+-hour interview, for which he or she would be paid ten dollars.[7]

The sampling scheme was therefore a stratified, geographic cluster approach. It was stratified to ensure the specific characteristics that we desired as major independent variables; it was geographically clustered to ensure a high probability that the target population would be encountered in the initial screening phase. Although the scheme was a departure from some "textbook" designs,

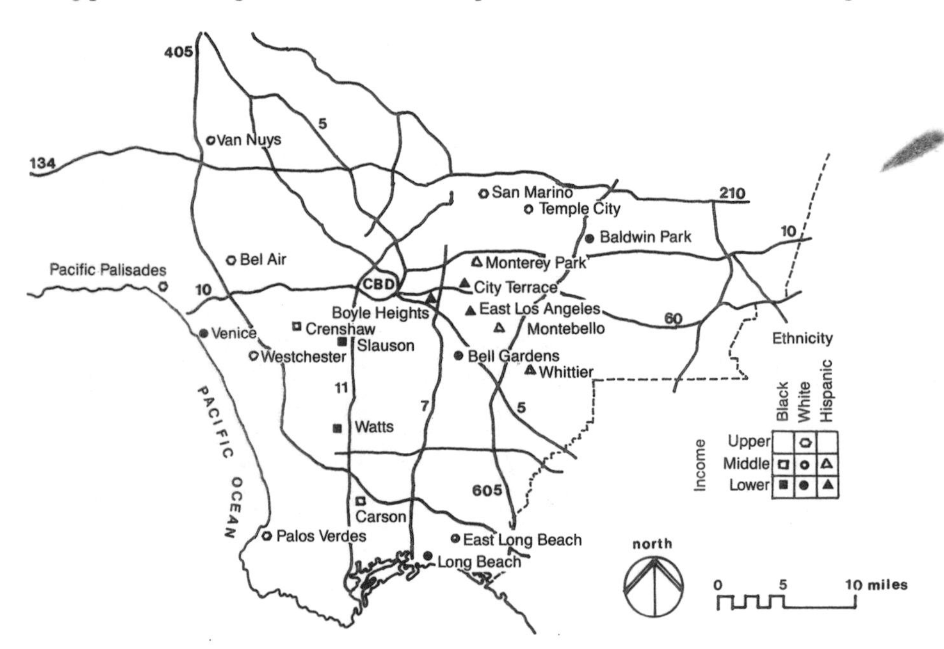

FIGURE 3.4. Location of the interview sites (identified by their social characteristics) in the Los Angeles area.

it recognized tight budget constraints and the realities of conducting field research on an increasingly reluctant public.[8]

PHYSICAL SETTINGS AND PERCEPTIONS: DIFFERENCES BY POPULATION GROUPS

Although these twenty-two areas varied greatly in their location and landscape, the basic features were often quite similar within a particular income group. For example, all upper-income settings shared low density and large lots, primarily single-family detached houses, consisting of an assortment of "ranch-style," "New England colonial," "Spanish revival," or "southern plantation" residential structures (Figure 3.5 and 3.6). These localities also had an abundant supply of private and public parks and open spaces; richly landscaped and tree-canopied streets; highly controlled and carefully planned commercial centers; and a well-maintained and well-serviced public environment.

The middle-income settings were all developed in the late 1940s or the 1950s. The houses were modest, built on small- to moderate-size lots arranged in a predominantly grid layout. One-story stucco units and medium-sized homes of ranch-style development were the common vernacular (Figures 3.7 through 3.10). All of these areas had some apartment structures located along the periphery or the main arterials (Figure 3.10). All of these areas were surrounded—and occasionally penetrated—by strip commercial and other high-intensity urban services.

FIGURE 3.5. A representative house in Pacific Palisades (upper-income white).

FIGURE 3.6. A representative house in San Marino (upper-income white).

The low-income settings were characterized by smaller street blocks and narrower streets. The housing stock was mixed—single-family duplexes, triplexes, and some apartments—and were generally in various stages of disrepair (Figures 3.11 through 3.14). These were often transitional areas, surrounded by commercial strips and, in some instances, by industrial sections. Major arterials or freeways had divided some of these areas into distinct sections. Undeveloped or vacant land, empty lots, and abandoned or boarded-up structures were com-

FIGURE 3.7. Houses of Van Nuys (middle-income white).

FIGURE 3.8. A residential street in Carson (middle-income black).

mon in these settings. Many of these areas were characterized by chain-link fences, graffiti on the walls, cars parked on lawns, high-tension lines, fast-food chains, gas stations, and auto-repair shops. Overall, the public environment was ill maintained and impoverished.

The verbal responses to our questions about the respondents' areas provided rich and diverse descriptions and evaluations, reflecting the qualitative and locational diversity of the settings and the variable concerns of the different social groups. Yet, some common themes recurred throughout all responses. Almost everyone seemed to acknowledge the existence of a social milieu while talking

FIGURE 3.9. Houses of Westchester (middle-income white).

FIGURE 3.10. Apartment complexes in Montebello (middle-income Hispanic).

FIGURE 3.11. A duplex in Boyle Heights (lower-income Hispanic).

FIGURE 3.12. A two-story residential structure in Venice (lower-income white).

FIGURE 3.13. Houses of City Terrace (lower-income Hispanic).

FIGURE 3.14. Protected dwelling space—an example from Bell Gardens (lower-income white).

about residential areas. Crime and personal or property safety appeared to be important aspects of the residential ambience. Smog was on most people's minds—a special sensitivity acquired, no doubt, from living in the Los Angeles area. Physical amenities and conveniences were mentioned by some respondents. The physical appearance of the area and its overall "atmosphere" (Milgram, 1970) were of concern to some people. The quality of the public services (schools, fire, police, and so on.) was another frequently mentioned aspect of the residential area.

The term *neighborhood* was carefully avoided in our initial questions about residential descriptions and evaluations. We had hoped that by asking people to talk about their *residential area*—a more general term—we would be able to elicit many different constructs of the residential area to assist us in thinking about design solutions. Contrary to our expectation, these open-ended responses did not yield any concrete notions of alternative concepts of the residential area. Occasionally, respondents referred to their residential area as a neighborhood, but more commonly they attempted to define their area using such social classifications as "upper-middle-class" or "laboring class" or such physical descriptors as "single-family homes" or "a complete residential area." But these are still too broad and general to guide design solutions. On the other hand, the common themes—social milieu, crime and safety, air quality, amenities, conveniences, appearance, atmosphere, and services—did shed some light on the existential reality that was perceived as the residential area.

What follows are discussions of responses to open-ended questions about how people characterized their area and what they liked and disliked about it. These are organized according to the major themes discussed above. The open-ended impressions and evaluations of the residential area are supplemented by responses to more structured questions about many of the same topics that were asked further along in the interview. These responses, although less wide-ranging, permit more precise comparisons among the different groups. They also allow the baseline comparison to be discussed later.

The Residential Area as a Social Milieu

It was clear from our interviews that most respondents recognized their area as fundamentally a social milieu. This social milieu may not necessarily have been referred to or recognized as a "neighborhood." Rather, it was likely to be described by categories of social class, race, or ethnicity. Furthermore, the descriptions of the social milieu sometimes took into account internal social dynamics, indicating whether that milieu was economically stable or in transition, socially homogeneous or heterogeneous, or was undergoing changes in ethnic composition. Some of the references to the social milieu were judgmental; that is, they concerned its suitability for rearing children; the rectitude of the people who lived in the area; and the conflicts and compatibilities between the respondent and the milieu. Such perceptions, to be sure, varied from locality to locality, and usually along the lines of income and race.

The upper-income respondents described their social milieu as "affluent" or "upper-middle class"; as being populated with like kinds of people who were also good neighbors; and as being the ideal social context in which to bring up children. Clearly, the upper-income people had been able to arrive at a place coveted by many and, naturally, were quite satisfied. Although a few wondered aloud about the implications of bringing up children in a socially homogeneous and sheltered community, no one advocated opening the door to other social

---

We are blissfully unaware of how the other half lives.

—SAN MARINO resident (upper-income white)

Upper middle class, hilly. Populated with some of the more expensive homes in L.A. I don't know what the people do. I don't talk with them.

—BEL-AIR resident (upper-income white)

No conflicts, everyone goes his own way; would not hesitate to help if help were needed. Are no clusters of ethnic groups; many professional people.

—PACIFIC PALISADES resident (upper-income white)

Upper middle class, large lots, well policed, broad streets, well-kept gardens

—SAN MARINO resident (upper-income white)

This immediate area is all single-family residential—they are working people. I think it's just a very ordinary suburb. It's not a young neighborhood—no swingers, that kind of thing. It's an older residential neighborhood

—VAN NUYS resident (middle-income white)

Middle-class, all-white neighborhood. Homes ranging from $20,000 to $30,000. It has nice schools. It's a nice area

—TEMPLE CITY resident (middle-income white)

It's middle to lower class. It probably goes lower than that in some places. It's mostly black and mostly residential.

—CRENSHAW resident (middle-income black)

Upper middle class or middle class, pride in their homes, nice children, respect for property rights, cordial to newcomers. . . . No poor people.

—WESTCHESTER resident (middle-income white)

It's clean. No neighbors that fight; the blocks are very close-knit. In the summer we look after each other's houses.

—VAN NUYS resident (middle-income white)

---

groups. As suggested by some respondents living in San Marino or Palos Verdes, theirs was a dream that had finally been realized.

The middle-income respondents appeared to be generally comfortable with their social environment. They described their fellow residents as "middle-class," "nice," and "looking out for each other's property." Nevertheless, a faint tone of apprehension and caution could be sensed in their comments. For example, the Westchester (white) residents appeared to be concerned about the changing nature of their area. The East Long Beach (white) residents resented

---

It's basically a middle-class neighborhood, with about average income. Some have better than average. They are mostly middle-aged with children.

—TEMPLE CITY resident (middle-income white)

It's not good to have older couples live with younger couples. They complain of the noise, parties, children

—MONTEBELLO resident (middle-income Hispanic)

Middle class. . . .Ethnic mix: Mexican-American, Armenian, few blacks, Latinos, Middle lower class.

—MONTEBELLO resident (middle-income Hispanic)

It's very close to cruising areas with a lot of car clubs, motorcycle gangs. Too many parties; neighbors fight.

—MONTEBELLO resident (middle-income Hispanic)

It's a mixture of racial groups and ages—mostly retired people and college-age people. Not typical middle-class suburbanites. It's not stucco palaces; the neighborhood has some character.

—VENICE resident (lower-income white)

It isn't as nice as twenty years ago—not as nice a class. . . . Now, in the last few years, people moving out. Mexican-Americans are moving in; no old people moving in.
—VENICE resident (lower-income white)

Some of the kids and adults are bad—drink, gamble, curse, immoral.
—WATTS resident (lower-income black)

Dope addicts, thugs, thieves; festering with bad influence.
—WATTS resident (lower-income black)

My street is peaceful and orderly and my neighbors are good people
—BOYLE HEIGHTS resident (lower-income Hispanic)

their nonconforming neighbors who did not keep up their property. The Van Nuys (white) respondents complained about bothersome teenagers.

Whereas the middle-income whites—as exemplified by the Westchester residents—appeared to be edgy about any racial change in their social milieu, their minority counterparts who lived in integrated areas welcomed racial heterogeneity, considering it a good thing. But like some of their white counterparts, the middle-income Hispanics were also concerned about the younger members of their communities—in particular, rowdy youths and reckless teenage drivers. In these communities, achieving harmony among different age groups may have been more of a problem than living next to a neighbor of a different race or color.

The social milieu became a major concern at the lower income level. The lower-income whites saw their social milieu as mainly unstable, as being populated by transient people without a real commitment to the area. They were suspicious of the increasing heterogeneity in both racial and life-cycle mix and resented deviations from established norms of behavior and lifestyle. People problems were most critical in the low-income black and Hispanic areas. Their comments conveyed a pervasive sense of fear, mistrust, anxiety, and alienation from their social milieu (Rainwater, 1966). Youth gang warfare and drug problems aggravated underlying tensions and were frequently mentioned.

Responses from the more structured portions of the interview revealed the same trends. Table 3.1 shows the respondents' perceptions of their respective social milieus by different population groups according to a global measure of desirability, and then according to particular measures chosen from the seven-point bipolar semantic differential scales.[9]

It is clear that the overall evaluation of the social milieu (i.e., "type of people" who live in the area) declined directly with income. Almost all of the upper-income group and more than two-thirds of the middle-income groups expressed a generally positive response, whereas the majority of the lower-income groups responded negatively.

This general pattern of declining evaluation with income held for most of the particular measures, except for "integrated-segregated"[10] and "friendly-

TABLE 3.1. Perception of Social Milieu by Population Groups

Social milieu considered by respondents as	Population groups						
	Upper white (n = 85)	Middle white (n = 80)	Middle Hispanic (n = 59)	Middle black (n = 86)	Lower white (n = 88)	Lower Hispanic (n = 55)	Lower black (n = 22)
	Global measure (%)						
At least ''somewhat desirable''	93.7	76.7	75.5	69.9	45.7	48.9	31.8
	Particular measures (%)						
At least ''slightly rich''	82.4	38.8	30.5	21.1	3.5	7.4	0.0
At least ''slightly high-status''	88.2	41.8	43.1	48.2	20.0	1.9	4.5
At least ''slightly integrated''	22.4	33.7	62.7	81.0	67.4	71.7	36.4
At least ''slightly friendly''	81.2	82.4	86.4	82.4	72.4	81.5	59.1
At least ''slightly personal''	60.0	56.3	57.6	52.9	37.9	45.1	22.7
At least ''slightly relaxed''	84.8	86.4	81.4	76.5	51.2	53.7	18.2
At least ''slightly talkative''	28.2	56.3	54.3	34.5	36.0	47.1	59.1

hostile.'' This last example is particularly interesting in the case of low-income groups that otherwise considered their neighbors undesirable. Perhaps, social desirability was not a direct function of sociability but a function of other dimensions of the social context, such as anomie, status, and stress. Also interesting is the comparative low evaluation on the ''personal-impersonal'' scale of the low-income groups, which contrasts with the popular belief that these groups have stronger ties with their immediate social milieu than do their middle- or upper-income counterparts (Fried and Gleicher, 1961; Fried, 1963; Michelson, 1970).

The Residential Area as a Haven

Safety of person and property is another important dimension of residential quality of life. People expect their residential area to offer safety and protection from the danger of the larger urban society. They would like to see it as a shelter, a refuge, but this is one expectation that remains largely unrealized for most urban residents; very few of our respondents sensed total immunity from crime, which was a matter of concern for all income groups, although little account of it was made in the neighborhood concept.

Even the upper-income respondents, who considered their environment a reasonably safe place, where one could walk at night without worrying about being mugged or assaulted, frequently mentioned burglary, theft, and other property crimes. Although they mentioned good police protection and emphasized that they were more safe than surrounding communities, the concern about safety was prevalent.

. . . its rather secluded; quite a number of criminals come out from the hills, the hillside.
—BEL-AIR resident (upper-income white)

Although we have had nine robberies, we have a good police department. They watch the house when we go on vacation.
—SAN MARINO resident (upper-income white)

There's really no good place to live in L.A. It's safer here than elsewhere, I guess, and the rent isn't bad.
—CRENSHAW resident (middle-income black)

When you speak of safety, I think of outside criminals. We have police surveillance, but the police department is in Venice, which is a long way away. Safety is policemen to me.
—WESTCHESTER resident (middle-income white)

It's not too bad. The kids throw cans and bottles on the lawn (the neighborhood wasn't like that before). They run their jalopies fast up and down the streets, and it's not safe for the kids.
—MONTEREY PARK resident (middle-income Hispanic)

I wouldn't walk down the street when it's dark. Don't get the police protection we should.
—BELL GARDENS resident (lower-income Hispanic)

All middle-income respondents were preoccupied with crime and police protection, even though only a few seem to have had first hand experience with robberies. Police protection was seen as adequate, although slow response time was occasionally a complaint among the middle-income whites. The reactions varied among locations. The inner-city blacks appeared to be more concerned than their suburban counterparts, underscoring a commonly held view, that crime is everywhere but is worse in the cities. The middle-income Hispanics, all from inner-city locations, considered property crime a problem but not a serious one. They were more concerned about violence, particularly that related to gang activities.

Crime and drugs together were the biggest complaint in all low-income locations. For the whites, this usually meant widespread robberies and use of narcotics by youths. In the Hispanic and black areas, fear of bodily assault and fights and killings between youth gangs were frequently mentioned along with robberies and narcotics. Crime in these areas was more likely to be in the form of personal assault than crimes against property.

Turning to the responses from the structured questions, we see these impressions confirmed: confidence in one's residential area as a safe and secure place declined rapidly with declining income (Table 3.2). Although the global measure of personal and property safety subsumed perceptions of threats other than crime to person and property, such as fire or traffic hazards, perceptions of crime clearly dominated these evaluations, as Table 3.3 reveals. Thus, it seems that safety of persons and property bolstered by adequate police protection, alarm systems, a dependable social milieu, and so on were the bases for the public's confidence in their residential area. And it was precisely the absence of such securities that made the low-income areas particularly vulnerable.

You can't sit at night with the door open 'cause they might shoot at you.
—CITY TERRACE resident (lower-income Hispanic)

Lots of thugs in the area. Too much thievery. People bad, fighting, killing.
—WATTS resident (lower-income black)

People out here are envious and want to take everything you get. They don't care about you.
—SLAUSON resident (lower-income black)

Guys break windows of cars that go by. Some areas are not well lighted. There's killings here between gangs. I don't feel very safe.
—CITY TERRACE resident (lower-income Hispanic)

We've had a very slight problem—personal danger, children twice been attacked by racial gangs. Dope problems.
—VENICE resident (lower-income white).

Rogues, thieves, fighting, gangs.
—WATTS resident (lower-income black)

TABLE 3.2. Perception of Personal and Property Safety by Population Groups

Population groups	Percentage reporting at least "somewhat safe" or better[a]	*n*
Upper-income white	94.0	85
Middle-income white	82.5	80
Middle-income Hispanic	73.7	59
Middle-income black	63.4	86
Lower-income white	47.0	88
Lower-income Hispanic	53.8	55
Lower-income black	9.1	22

[a]That is, including "safe" and "extremely safe" ratings.

Residential-Area Air Quality

Like crime and safety, air quality was not a concern when the neighborhood principles were formulated. Even today, smog and air quality may not be a major factor in judging the residential quality of life outside major metropolitan areas. In Los Angeles, air quality is known to have significant effects on property values and on people's health, mood, and leisure habits. It is seen as a continuing

TABLE 3.3. Most Frequently Given Explanations for Evaluations of Personal and Property Safety

Explanations[a] most commonly given by: Those who considered their area at least "somewhat safe" or better	Those who considered their area "in between" (i.e., neither safe nor unsafe) or worse
1. Property safety	1. (Lack of) property safety
2. Police patrol	2. (Lack of) personal safety
3. Personal safety	3. House safety
4. Alarm system/other precautions	4. Type of people
5. Type of people	5. Traffic safety
6. Traffic safety	6. (Lack of) police patrol
7. House safety	7. Street lighting
8. { Location	8. Location
8. { Neighbor vigilance	9. Fire patrol
9. Fire patrol	10. Alarm system
10. Fire safety	

[a]Ranked according to frequency of mention in each category of response.

I smell it. I see it (doesn't make my eyes burn, though). I think about what it does to my children's lungs.

—SAN MARINO resident (upper-income white)

It is practically smog-free. The temperature in summer is cooler than the central city. It is warmer in the winter. There is always a breeze in the afternoon.

—WESTCHESTER RESIDENT (middle-income white)

We're in the San Gabriel Valley, in a pocket, and the smog just lies in here.

—SAN MARINO resident (upper-income white)

I wish someone would study the air here, 'cause I swear we have the best in the county.

—WESTCHESTER resident (middle-income white)

Air-quality problems from the refinery. It stinks, bad odor.

—CARSON resident (middle-income black)

---

environmental menace that most Agelenos would like to see eliminated. Of the twenty-two locations, five (comprising white areas of all income classes)—Pacific Palisades, Venice, Palos Verdes, Long Beach, and East Long Beach—are coastal and therefore enjoy clean air. Naturally, respondents from these areas were quite happy with the air they breathed and frequently mentioned this fact as a positive aspect of these settings.

Almost all nonwhite locations—regardless of their income class—are inland and less fortunate in their air quality. Many are located in the "smog belts" of the San Gabriel Valley or south-central Los Angeles.

In general, the upper- and middle-income respondents appeared to be more concerned about the quality of the air than their lower-income counterparts. Those who enjoyed clean air by virtue of their coastal location made a special note of it, comparing their lot with that of the rest. Those who did not, appeared distressed—for example, the upper-income residents of San Marino (white)—complained about the air quality of the San Gabriel Valley. The Bel-Air (upper-income white) residents were uncomfortable about living near the San Diego Freeway and complained about fumes and pollution. The East Long Beach (middle-income white) residents, despite their proximity to the coast, were unhappy about the presence of a nearby utilities plant. Additionally, they found the fumes and jet-fuel emissions from air traffic of the Long Beach airport annoying. Similarly, the Carson (middle-income black) residents complained about the fumes from nearby refineries and the newly developing industrial park. Like the San Marino residents, respondents from other San Gabriel Valley locations—Temple City (middle-income white), Monterey Park (middle-income Hispanic), and Montebello (middle-income Hispanic)—were also worried about the smog.

Despite overall poor air quality, the lower-income respondents—with the exception of Venice (white) and Long Beach (also white)—did not talk much about smog. Even though they were dissatisfied with the air—as we will see elsewhere—it clearly ranked low in their order of concerns and priorities. As

Not very often do we get a clear blue sky. It's smoggy about 88 percent of the time. We can't see the hills about half a mile away. Your eyes burn and you get sluggish and tired.

—MONTEREY PARK resident (middle-income Hispanic)

You can smell it—smell the dirt. Can't see the beautiful mountains. Sky is yellow. You can feel it—lungs hurt.

—VAN NUYS resident (middle-income white)

Well, the entire city is smoggy.

—SLAUSON resident (lower-income black)

We have had pleasant days in the past year. But we get quite a bit of smog from the freeway.

—BELL GARDENS resident (lower-income white)

People have to get around, and there is no fast way like the Metro in Mexico, so you have to use the car or bus.

—BOYLE HEIGHTS resident (lower-income Hispanic)

discussed previously, they were more preoccupied with crime, narcotics, drugs, and violence. Still, occasional complaints could be heard. For example, the Boyle Heights residents, who were encircled by a network of freeways and interchanges, emphasized the freeways as the immediate source of air pollution.

Turning to Table 3.4 we see that ambient air quality also appeared to be directly related to income class. With the exception of the upper-income whites, the majority of all other respondents did not consider their air quality even "somewhat clean." This finding is intriguing, as some of the middle- and lower-income white areas (two out of four in each case) are located near the coast and are therefore likely to enjoy cleaner air (Figure 3.4). However, both the coastal location and the data obtained from air-quality monitoring stations may belie the local variations (and therefore, the local perceptions of air quality), which can be affected by the presence of industries, refineries, freeways, parking lots, and other land uses known to be major stationary sources of contaminants (such as gas stations, cleaners, and so on). When we asked our respondents to explain their perceptions of air quality, they mentioned not only the conditions of air

TABLE 3.4. Perception of Air Quality by Population Groups

Population groups	Percentage reporting at least "somewhat clean" air or better	n
Upper-income white	61.9	85
Middle-income white	25.6	80
Middle-income Hispanic	19.6	59
Middle-income black	27.5	86
Lower-income white	20.2	88
Lower-income Hispanic	16.0	55
Lower-income black	18.2	22

TABLE 3.5. Most Frequently Given Explanations for Evaluations of Air Quality

Explanation[a] most commonly given by:	
Those who considered their air at least ''somewhat clean'' or better	Those who considered their air ''in between'' (i.e., neither clean nor smoggy) or worse
1. (Presence of) sea breeze, wind, or other climatic or geographic factors	1. Physical discomfort
2. (Absence of) haze or lack of visibility	2. Haze or lack of visibility
3. (Better) compared with another place/location	3. Industry, freeways, airplanes, and other such sources
4. (Absence of) physical discomfort	4. (Worse) compared with another place/location
5. (Absence of) odor	5. (Absence of) sea breeze, wind, and other climatic and geographic factors
	6. Odor

[a]Ranked according to frequency of mention in each category of response.

pollution (or lack of it) but also the factors that they perceived to be contributing to the problem (or its alleviation), as shown in Table 3.5.

The Residential Area as an Amenity Resource

A common theme in descriptions of residential areas pertained to amenities and conveniences. Amenities are those qualities of the environment that make life more pleasant and enjoyable, such as having a park nearby, having easy access to the beach, being close to a museum or other cultural facilities, having a breathtaking view from one's living-room window, or having quiet and privacy. Conveniences are easy availability of services and facilities for everyday needs;

Chamber of Commerce is careful about who they let in—like not allowing the "Jack in the Box"—keep everything in line; have trash pickup drives, and so on. Close to many things—you can drive or walk to most anything—beach, ski, anywhere, anything.

—PACIFIC PALISADES resident (upper-income white)

It's located reasonably conveniently to the many different areas in L.A.—40 minutes or less to downtown L.A. or the valley, or the beach. It would be a fairly long walk to a shopping area. The children will have to be driven to school.

—BEL-AIR resident (upper-income white)

Stores are pretty close. We get a bus a couple of minutes away. It's got everything we need right in this area, pretty close.

—WHITTIER resident (middle-income Hispanic)

FIGURE 3.15. Recreational facilities and amenities in Pacific Palisades (upper-income white).

good public transportation; good shopping areas nearby; and easy access to regional entertainment facilities or schools for children.

As might be expected, the upper-income areas were described by their residents as high-amenity settings. All these locations had an abundant supply of private and public parks and open spaces, as well as a well-maintained and well-serviced public environment. They were also quiet and private. Pacific Palisades and Palos Verdes overlook the ocean and are close to beaches; Bel-Air and San Marino are close to such cultural facilities as universities and museums. Although the residents enjoyed these amenities (Figures 3.15 and 3.16), they were less enthusiastic about their locational conveniences. The very qualities—low

---

No public transportation to speak of. No large grocery stores. No doctors, druggist, little park but nothing in it for small children to play with.

—BALDWIN PARK resident (lower-income white)

Small residential area; facilities are within immediate area. Shopping, transportation, city utilities; with continuous improvement in educational buildings, highways, and busing.

—MONTEREY PARK resident (middle-income Hispanic)

It is a pleasant community. You feel safe. The neighbors are really friendly. The kids like it. We just like it here. It is safe. It is clean. The schools are good. Everything is close by, shopping, buses; it's a good location. You can get along with the neighbors; you can depend on them.

—WHITTIER resident (middle-income Hispanic)

FIGURE 3.16. Recreational facilities and amenities in Pacific Palisades (upper-income white).

density, exclusive land use, suburban or semirural locations—that make these areas high-amenity resources also make them less conveniently situated with respect to shopping, schools, entertainment, and similar activity settings. Most Palos Verdes residents, for example, were acutely aware of the trade-off involved in the enjoyment of these amenities: they were far from the city and isolated from many regional and cultural facilities and other urban services, which were difficult to get to without a car.

The middle-income respondents tended to emphasize the conveniences in their area more than the amenities. Occasionally, they spoke of quiet as an amenity, but many of them also complained about excessive noise in the area—a disamenity. Overall, most middle-income respondents considered their location convenient to schools, shopping, and various regional facilities. The Westchester residents, for example, enjoyed their proximity to a major shopping and business area. The East Long Beach residents spoke of their proximity to three freeways—which increased their access potential. Easy access to schools and shopping were particularly emphasized by the Temple City residents. The Carson residents, however, were not particularly happy about their locational opportunities and complained about inadequate access to shopping, schools, parks, and recreation (see Figures 3.17 through 3.19).

At the lower-income level, references to amenities were nonexistent, and perception of conveniences was mixed. For example, the Venice residents, despite all their problems, acknowledged the convenient location of stores, but the residents of Baldwin Park complained about the high costs of goods (resulting from the nonavailability of large retail outlets) and the inadequacy of storm drains, sidewalks, the sewage system, and public transportation. The residents of Boyle Heights talked about the lack of street lighting and about the noise and the filth in the area. At the same time, they liked being close to churches and a major hospital. The City Terrace residents similarly considered themselves conveniently located, despite their many social problems. The East Los Angeles residents wished that there were an adequate shopping center nearby, but otherwise they liked being near churches, clinics, and a major hospital.

In summary, we see from Table 3.6 that a substantial majority of all but one of the population groups (lower black) considered their areas at least somewhat convenient. This finding may strike an aficionado of Los Angeles as a tribute to its efficient urban form. The spatial diffusion of land use and activities that is considered the root of many environmental problems in the Los Angeles region appears to have produced a highly redundant and accessible urban form throughout most of the area. Indeed, as we examined the median travel times for six different opportunity items reported by the population groups, and as we compared these figures with the standards recommended by *Planning the Neighborhood* (APHA, 1960) and a more recent handbook of planning and design standards (deChiara and Koppelman, 1975), it became evident that all of these figures were well within the upper bounds of the prescribed standards.

FIGURE 3.17. Leimart Park and adjoining shopping in the Crenshaw district (middle-income black).

FIGURE 3.18. A quiet residential street in East Long Beach (middle-income white).

FIGURE 3.19. A major shopping district in Westchester (middle-income white).

TABLE 3.6. Accessibility of Different Facilities, Perception of Overall Convenience by Population Groups, and Some Related Planning Standards

	Median travel time[a] in minutes to:						
Population groups	Work	Friends and relatives	Schools for children	Culture and entertainment	Shopping	Recreation	Percentage who considered their area at least "slightly convenient" or better
Upper white (n = 85)	16.6	16.5	8.1	24.4	8.0	12.1	75.0
Middle white (n = 80)	17.7	13.1	8.0	24.0	8.0	13.5	85.0
Middle Hispanic (n = 59)	17.0	8.3	7.6	18.6	7.7	12.4	86.3
Middle black (n = 86)	18.6	13.5	11.3	24.1	7.8	17.8	74.1
Lower white (n = 88)	16.4	12.0	11.5	24.4	8.0	11.8	65.5
Lower Hispanic (n = 55)	24.0	17.9	12.3	24.9	13.5	51.7	77.8
Lower black (n = 22)	28.8	8.8	8.1	23.8	15.0	24.2	40.9
Related planning standards are shown below							
Planning the neighborhood (1960)	20–30		< 15[b] < 20[c] 15–25[d] 20–30[e]	20–30	< 20	< 20[f] 30–60[g]	
deChiara and Koppelman (1975)	60		20–30	60–90	< 20	< 30[f] 45–60[g]	

[a]Based on the "access" cards filled out by the respondents in "playing" the trade-off game. See Appendix I; [b]Nursery school and kindergarten; [c]Elementary school; [d]Junior high school; [e]Senior high school; [f]Local recreation;[g]Major outdoor recreation.

Still, there is more than a suggestion that the low-income minorities suffered some comparative locational disadvantages. They appeared to be further removed from their places of work, shopping, and recreational facilities than were the rest. Apparently, even the highly redundant and accessible urban form of Los Angeles failed to provide the same level of conveniences and opportunities to these areas.

The Residential Area as a Physical Place

A sense of place is based on the overall "look" and "feel" of an area (Lynch, 1976). As might be expected, the upper-income respondents spoke highly of the appearance and the atmosphere of their environments. They viewed their areas as unique, beautiful, and rural. The Pacific Palisades residents talked about the "small-bedroom-community" atmosphere of the area, the controlled and exclusionary land use, and the abundance of vegetation and landscaping. The Bel-Air residents noted the rural nature of the area, its proximity to undeveloped hillside, and its absence of sidewalks. The residents of Palos Verdes were highly appreciative of the area's natural and manmade beauty and its rural atmosphere (see Figures 3.20 and 3.21).

The physical layout of the area and the housing rarely attracted much comment from the middle-income respondents: it was better than adequate, but not worthy of praise. It seemed to meet closely the residents' expectations so that their attention was directed to nonphysical concerns. Still, there was occasional reference to the physical form of the area. For example, the Van Nuys apartment-

FIGURE 3.20. Tree-lined streets in Bel-Air, an upper-income area.

FIGURE 3.21. Tree-lined streets in San Marino, an upper-income area.

dwellers complained about the nearby commercial land use and the resulting heavy traffic. Confusing street patterns were a source of minor complaint for the Crenshaw residents (see Figures 3.22 through 3.25).

It's beautiful; the trees, and home restrictions to pastel shades for houses, and tile roofs; the terrain—the hills next to the sea. The general atmosphere of winding streets and curbside mail boxes.

—PALOS VERDES resident (upper-income white)

I would say it's country-like atmosphere: many trees, flowers, lawns, and parkways, bridle paths, bicycle trails. I don't know what else you want—no mess, rather large and pretentious, very picturesque.

—PALOS VERDES resident (upper-income white)

It looks nice here. It seems well planned. It is centrally located.

—MONTEBELLO resident (middle-income Hispanic)

(We are) surrounded by bars, liquor stores, motorcycles, police, cars, ambulances, fire engines, trucks, buses—fumes from the buses—and heavily congested traffic. It's sometimes so oppressive I can hardly breathe.

—VAN NUYS resident (middle-income white)

Nice area: quiet, peaceful atmosphere—close to country atmosphere, L.A. is much too congested.

—CARSON resident (middle-income black)

FIGURE 3.22. Heavy traffic bordering residential areas—Westchester (middle-income white).

FIGURE 3.23. Mixed land use in Van Nuys.

FIGURE 3.24. A mural adds life to an otherwise featureless residential street in Crenshaw (middle-income black).

In the low-income areas, the appearance and atmosphere of the area were of more than casual concern. The respondents frequently used the word *dirty* to describe these areas. Venice residents complained about unfenced and unleashed dogs, narrow streets, small lots, high density, and crowding. Bell Gardens respondents spoke of the poor maintenance of rental homes. The Hispanic respondents from Boyle Heights and City Terrace also complained about dirty or unpaved streets and inadequate street lighting. But these were not concerns for

---

It's mostly apartments on a main street. A little liquor store you walk to, a drug store and a pharmacy. I don't think it's that residential an area; it's mostly apartments. It's noisy. The only thing I like about it is that it is convenient. Other than that, it is blah.

—VAN NUYS resident (middle-income white)

Nice, quiet neighborhood; middle-class neighborhood. I used to live in another area, and on weekends, I would hear police sirens, ambulances, etc., but that does not happen here.

—CARSON resident (middle-income black)

Dirty. Awful lot of dirt here all the time. It blows in the windows. It could be beautiful—some nice yards and people, but I don't know them.

—BELL GARDENS resident (middle-income white)

Most of the time, the streets are dirty and dusty, so it makes the air unpure to breathe.

—EAST LOS ANGELES resident (lower-income Hispanic)

FIGURE 3.25. Even oil derricks can be a part of a residential landscape—an example from Carson (middle-income black).

their black counterparts. They seemed to be more preoccupied with income, rents, and other pressing social and livability problems (see Figures 3.26–3.30).

The structured questions addressed themselves to some other facets of the residential area as a physical place. In general, people were more satisfied with their dwelling space than with their larger residential space. Only the upper-income white groups claimed that their residential area was spacious. In Table 3.7, we present evaluations of the physical place in terms of space, privacy, comfort, aesthetics, ambience, and maintenance. Clearly, there was a difference between income groups, with higher-income groups generally having more of the desirable features. Still, among the lower-income Hispanics, a surprisingly high percentage of persons reported a positive response to the various residential attributes. In other respects, the pattern was not as clear-cut as it had been for some of the other considerations.

Noisy kids, not enough space; dirty, oily parking area; too many thieves.

—WATTS resident (lower-income black)

This isn't a good place. I live here because I have to. We are all poor out here and nothing is ever good about our plight. Our conditions are deplorable. We must organize, but it is difficult to appeal to the poor, wretched slaves. We are slaves, you know, and the only way a slave can change his condition is to revolt and rage undammed. The cost of living in the slave quarters helps perpetuate our condition. Everything is beyond our purchasing power. The gangs who are doing the killing—slaves killing slaves because they are afraid to kill the enemy. They are frustrated and take their hostilities out on each other.

—WATTS resident (lower-income black)

FIGURE 3.26. Graffiti decorate a corner market in City Terrace (low-income Hispanic).

FIGURE 3.27. One of the many murals that adorn Venice (lower-income white)—a feature that was hardly mentioned by the residents.

TABLE 3.7. Perceptions of the Physical Place by Population Groups

	Population groups (%)						
Area considered by respondents as:	Upper white (n = 85)	Middle white (n = 80)	Middle Hispanic (n = 59)	Middle black (n = 86)	Lower white (n = 88)	Lower Hispanic (n = 55)	Lower black (n = 22)
At least ''slightly spacious''	77.6	32.5	37.9	39.3	14.9	38.9	31.8
At least ''slightly private''	96.5	76.2	67.8	63.5	40.7	40.7	13.6
At least ''slightly beautiful''	97.6	65.0	76.3	81.0	32.2	44.4	22.7
At least ''slightly cared-for''	95.3	82.5	76.3	86.9	34.5	64.8	50.0
At least ''slightly quiet''	69.4	38.8	47.5	57.6	36.8	60.0	22.7
At least ''slightly natural''	47.1	12.5	13.6	13.3	9.2	14.8	27.3
At least ''slightly lasting''	64.7	41.3	33.9	27.4	31.9	76.9	77.3
At least ''slightly comfortable''	97.6	91.2	94.9	88.1	70.1	81.5	40.9
At least ''slightly clean''	95.3	83.8	86.3	74.1	65.5	77.8	40.9
''Just enough dwelling space'' or better[a]	90.4	72.0	74.1	75.3	67.1	76.0	45.5

[a]From the ''adequacy of dwelling space'' card filled out by the respondents in playing the trade-off game. See Appendix I. Note, however, that although this item pertained specifically to evaluations of the dwelling space, the first item in the list, derived from the bipolar semantic differential scale, referred to the residential area at large. Thus, these two measures were quite different in scope.

FIGURE 3.28. Lack of maintenance and care—a street in Boyle Heights (lower-income Hispanic).

Linked to the sense of the physical place is the concept of a community. The "eclipse" and decline of place and community has been a major topic of discussion among the students of urban social processes (Suttles, 1975; Webber, 1964; Fischer, Jackson, Stueve, Garson, Jones, and Baldassard, 1977). It will be recalled that one of the contextual values of the neighborhood unit concept was to restore, maintain, and create a sense of community. Reference to community, however, was infrequent in our residential descriptions, although it was mentioned consistently by respondents from three localities: San Marino (upper-income white), Whittier (middle-income Hispanic), and Venice (lower-income white). In all three cases, the city or community, rather than the immediate neighborhood, appeared to be the frame of reference. In the case of San Marino, this response could be attributed to the exclusivity of the area, which was maintained by a self-selected group (self-selected not only by virtue of income but also by virtue of social values) with a shared political affiliation (a pronounced emphasis on conservatism, with frequent reference to the John Birch Society). Thus, moving to San Marino was like joining an exclusive "country club," and the affiliation with the entire community was more important than was one's immediate residential area. Another possibility of San Marino's preeminence as a "community" was its age. It was one of the older communities, built prior to the 1940s as a prestigious community, whose status and attraction had remained undiminished since its beginning.

FIGURE 3.29. Broken sidewalk and litter—Venice (lower-income white).

Similarly, the history of the community might have endowed Whittier and Venice with this particular quality. Whittier was built in the 1940s and was older than most of the other middle-income neighborhoods in which we interviewed. It also had some distinctive physical features—a hill and boundaries defined by freeways—that may have contributed to its sense of community. The symbolic value of being the hometown of a former U.S. president may have added some weight. Venice, of course, is a legend in Southern California, built at the turn of

FIGURE 3.30. Vacant, waste areas in the Baldwin Park (lower-income white) landscape.

the century as a coastal resort town, complete with canals and gondolas. Most of the artifacts of this curious era of a California romanticism with a European tradition have disappeared with time; the canals still remain, but only in an impoverished state. And so does the name "Venice." Even though the original physical community had been more or less obliterated by steady encroachment on its fringes by new developments, the lingering aura of the bygone days still made its designation as a "community" credible.

In conclusion, the sense of community as expressed in these residential descriptions amounted to something larger than the immediate residential setting. Where mentioned, it appeared to exist as serendipity or tradition rather than as a result of public policy. But more significant, in a large majority of cases, the community was *not* mentioned. The appearance and "atmosphere" of the physical place were more consistently recognized by most groups.

The Residential Area as a Conduit of Public Services

It is apparent from some of these descriptions that the residential area can also be seen as a conduit of public services: schools, police, fire, street cleaning, garbage collection, and so on. Once again, the upper-income whites seemed to be content with the level of governmental services and the school system. But in those years immediately preceding Proposition 13, high property taxes were often mentioned. Although the quality of public services was appreciated, there was also the general feeling that these were heavily paid for through taxes.

The middle-income whites were also satisfied with their schools, police, and fire services, but occasional complaints about high taxes could also be heard. Among the middle-income blacks, opinions about transportation and police were mixed. The suburban Carson residents were unhappy about transportation; inner-city Crenshaw residents were disgruntled about the police. Most middle-income Hispanics were pleased with their schools and public maintenance services; however, only the Monterey Park residents reported satisfaction with police.

The record of public services was much less satisfactory in the low-income areas. For example, the Baldwin Park residents complained about inadequate

---

If we have to call the police, they get here within a reasonable amount of time—oh, ten minutes.

—WESTCHESTER resident (middle-income white)

The lights in the street do not work very well. When it rains, the lights go off—even if it does not rain. Markets and transportation are far. We have to walk across the bridge. My church is also far. The streets are not very clean. We don't have a health center. There is one close, but they say we don't belong to it. We have to go to the General Hospital.

—CITY TERRACE resident (lower-income Hispanic)

In the immediate vicinity there is much opportunity for education, i.e., Dominguez Hills and Cal State Long Beach; much in the line of recreation, schools and parks; security services rendered by the police should be commended; industrial location factor, i.e., industrial park—hence much opportunity for unemployed. But too far from place of worship; my acquaintances are too far away; shopping areas are not near, i.e., choice clothing stores.
—CARSON resident (middle-income black)

No recreation for children or adults. Recreation provided is too far away, and children must have transporation to get there. A park of sorts is needed in this area.
—CARSON resident (middle-income black)

I have good transportation. We have churches nearby. We have well-lighted grounds and streets. Paved streets. We have social activities available in our residential area. Our area is clean and attractive.
—EAST LOS ANGELES resident (lower-income Hispanic)

storm drains, sewage systems, and public transporation. The residents of City Terrace also complained about the lack of public transporation and, in addition, about the bothersome presence of police. Only the residents of Boyle Heights and East Los Angeles expressed satisfaction with public transporation, health services, and maintenance of the public environment.

DIFFERENCES IN PERCEPTIONS BY STAGES IN THE FAMILY CYCLE

So far, we have presented our findings largely in terms of differences between different population groups. No attempt was made to sort them out by stages in family cycle because of the complexity and the tedium involved in the analysis and reporting of open-ended responses at such disaggregated levels. Instead, we focused on the pertinent scaled-response items that were amenable to some controlled analysis, such as the multiple classification analysis. Here we discuss some of the general findings related to stages in family cycle. In the case of most of the scaled responses, we were able to examine the adjusted means for different stages in the family cycle, once the effects of the population groups were controlled through a multiclassification analysis.[11] Generally, we found that, although the effects of the population group were almost always statistically significant, this was rarely the case for the stages in the family cycle. Nevertheless, it is important to look at such differences where they occurred and to discuss their implications for planning and design.

With regard to "social milieu," we found significant differences within stages in the family cycle in the cases of the "integrated-segregated" and the "friendly-hostile" scales. The elderly were likely to perceive their social milieu as more integrated and friendly than the other two groups in the family cycle,[12] but no such significant difference could be noted in the cases of perception of safety,[13] air quality,[14] or convenience, once the effects of population groups were controlled. However, significant differences were noted in perceptions of spaciousness, adequacy of dwelling space, and the comfortableness of the residential area. Typically, the elderly were more generous in their evaluations, whereas younger households, with or without children, tended to be more stringent. As noted previously, the elderly appeared to be a bit easier to satisfy, a reaction reflecting, perhaps, an age-related tempering of expectation.[15] The other groups, because of their children or their lifestyle expectations, appeared to be more demanding in their requirements for their immediate setting.

SATISFACTION AND PRIORITIES

In the preceding discussion, we have tried to develop some of the main themes that underlay people's impressions of their residential environment. In the process, we have also tried to point out the significant differences in those concerns between different social groups. What we have not explored so far is which of these concerns were more important than others, and to whom, and how satisfied different groups were with different aspects of their residential environment. In this section, we explore these issues by drawing on data obtained in a different part of the interview.[16] In it, the respondents were engaged in playing a trade-off game, as part of which they were required to indicate their satisfactions and priorities with a selected set of eleven residential attributes. Six of these attributes concerned the accessibility of significant places: work, schools, friends and relatives, cultural and recreational facilities, shopping, and recreation. The other attributes represented significant aspects of the overall residential ambience: air quality, dwelling space, density, personal and property safety, and the type of people who lived in the area. Although these items did not cover everything implicit in the open-ended response, they corresponded to most of these general concerns.

The essential differences between the seven population groups are shown in Figure 3.31 in a graphic representation of what can be called profiles of satisfaction. The eleven spokes of the wheel represent the eleven attributes described above. Each spoke shows the corresponding mean satisfaction score for a particular group on a seven-point scale, where a score of 1 means "very unsatisfying" and a score of 7 means "very satisfying." The shaded area represents a scale value of 4 or lower, corresponding to the range of ratings from "neither satisfying nor unsatisfying" to "very unsatisfying." We assumed here that a score of 4 was the threshold of dissatisfaction.

The change in the satisfaction profile from the low- to the high-income levels is quite obvious. The upper-income group not only had higher levels of

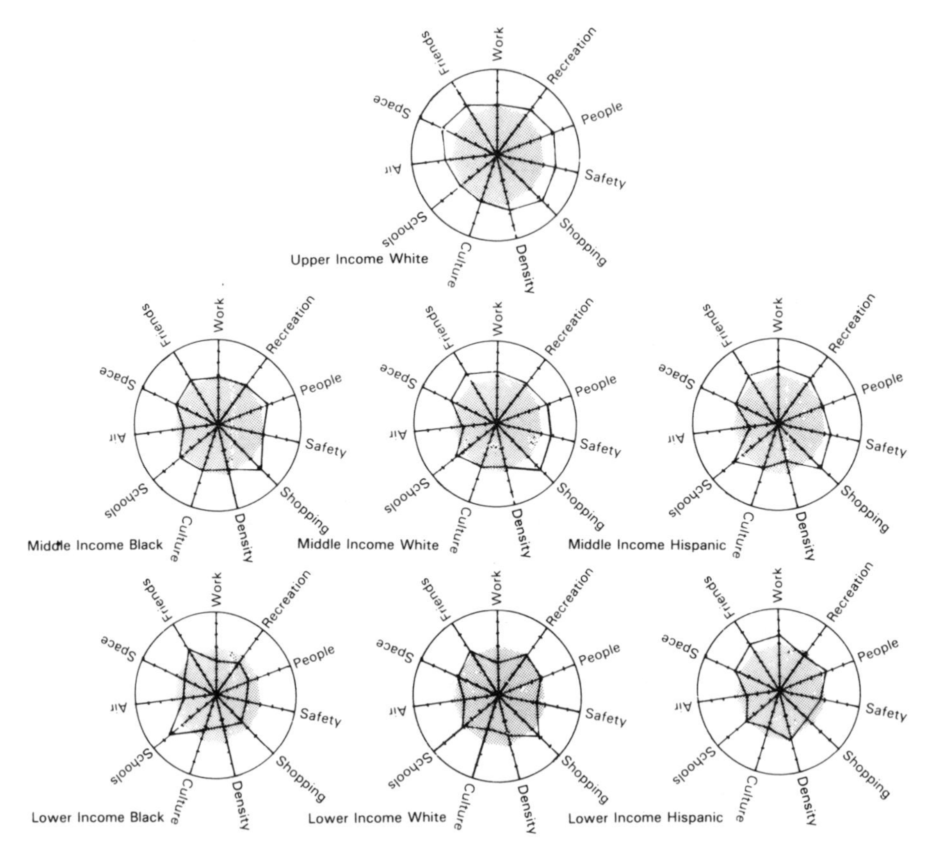

FIGURE 3.31. Satisfaction profiles by population groups (shaded area indicates dissatisfaction).

satisfaction (never dropping into the shaded area of dissatisfaction) but also had the most rounded profile, indicating a roughly equivalent degree of satisfaction with *all* aspects of the residential environment (perhaps with the exception of "access to schools"). The middle-income groups not only showed lower levels of satisfaction but also revealed a less rounded profile. And the lower-income groups showed the least satisfaction with most attributes, as well as the least rounded profile. The only attribute satisfactory to all groups was "access to friends and relatives." Generally, it seemed that the middle- and lower-income residents were more satisfied with their access levels than with their ambient qualities. In the case of the upper-income groups, there was slightly less satisfaction with access, but it must have been only a slight dissatisfaction, because, after all, the choice of living in an affluent neighborhood—characterized by low density and a long commute from the center—had been a deliberate decision for these people. They had consciously made trade-offs between distance and ambience. Indeed, it seemed that almost all groups were willing to make that kind of trade-off, but whereas the rich had been able to afford it, the others could only aspire to it.

In another part of the trade-off game, the respondents were asked to allocate twenty-five poker chips among the eleven attributes as a way of indicating their relative importance.[17] Tables 3.8 and 3.9 show the ranking of the attributes according to their adjusted mean scores by different population groups and stages in family cycle. That is, in both of these figures, the attributes are ranked according to the average number of chips allocated to them by different groups in the trade-off-game (see Appendix I).

It appears that ambience attributes, generally, ranked higher than access attributes, but there were some major exceptions. Access to school ranked high for households with children, and also for some minority groups, except the lower-income Hispanics. Although our data were collected before the Los Angeles school-desegregation controversy of the late 1970s came to a head, these concerns had been smoldering for quite some time, and it is possible that the priority given to access to schools was symptomatic of these concerns. The only other access attribute that appears to have had some higher order priority was access to shopping. Access to work did not rate high, although as Figure 3.31 shows, low-income Hispanic and black groups were dissatisfied with their present level of access to work opportunities, which was probably indicative of the adverse impacts of the growing suburbanization of jobs on the entrenched inner-city minority workers.

These two cases aside, it is clear from Tables 3.8 and 3.9 that ambience attributes received top priority for all groups, albeit in different order. It is, however, revealing to see how the relative ranking of these various ambience attributes varied between groups and corresponded to the intergroup differences in the open-ended responses discussed previously.

It is particularly interesting to note that air quality was considered the highest ranking ambience attribute by all white groups (upper, middle, and low), but not by the nonwhites. Indeed, it ranked low for blacks. Safety was an attribute of highest priority among all nonwhite groups, except the low-income blacks, who considered their immediate social milieu even more important than safety. Of course, for residents of impoverished black ghettos, the "type of people" who lived in the area determined "personal and property safety."

Perhaps these rankings reflect the different need hierarchies of different social groups—in the Maslovian sense (Maslow, 1962)—shaped by their current deficiencies, deprivations, and vulnerabilities in the immediate residential environment. In any event, the rankings corroborated what we had already gleaned from the open-ended responses and what other data sources had displayed previously.

TABLE 3.8. Priorities in the Residential Environment by Population Groups (Mean Chip-Allocation Score[a])

Upper-income white ($n = 85$)		Middle-income white ($n = 80$)		Middle-income Hispanic ($n = 59$)		Middle-income black ($n = 86$)		Lower-income white ($n = 88$)		Lower-income Hispanic ($n = 55$)		Lower-income black ($n = 22$)	
Air	3.5	Air	3.8	Air	3.3	Safety	3.6	Air	3.7	Safety	3.8	People	5.9
Safety	3.3	Safety	3.0	Safety	3.3	People	2.9	Safety	3.0	Air	3.0	Safety	4.6
Space	3.0	Space	2.7	Schools	2.8	Space	2.8	Space	2.7	*Friends*	2.7	*Shopping*	3.1
People	2.5	People	2.3	*Recreation*	2.3	*Schools*	2.6	People	2.5	*Shopping*	2.7	*Schools*	2.7
Shopping	2.3	*Shopping*	2.2	*Shopping*	2.3	Air	2.5	*Schools*	2.3	Space	2.5	Density	2.6
Work	2.0	*Work*	2.1	Space	2.3	*Shopping*	2.5	*Shopping*	2.3	*Schools*	2.4	Space	2.5
Schools	1.9	*Schools*	2.1	*Work*	2.1	*Friends*	1.9	*Friends*	2.0	People	2.3	Air	1.1
Friends	1.8	*Friends*	2.0	Friends	2.0	*Work*	1.8	*Recreation*	1.9	*Culture*	1.8	*Work*	0.8
Culture	1.7	Density	1.7	People	2.0	Density	1.7	Density	1.6	*Work*	1.3	*Friends*	0.7
Density	1.7	*Recreation*	1.7	*Culture*	1.4	*Culture*	1.6	*Culture*	1.4	*Recreation*	1.2	*Culture*	0.6
Recreation	1.7	*Culture*	1.4	Density	1.0	*Recreation*	1.6	*Work*	1.3	Density	0.9	*Recreation*	0.4

[a]Adjusted for the stages in family cycle (from a multiple-classification analysis of initial chip allocation involving a two-way analysis of variance in which "population groups" and "stages in family cycle" were introduced simultaneously; (the values in each column add up to 25). Italicized attributes are access-related.

TABLE 3.9. Priorities in the Residential Environment by Stages in Family Cycle (Mean Chip-Allocation Score[a])

Households with children (n = 256)		Households without children (n = 117)		Elderly (n = 102)	
Safety	3.2	Air	3.6	Safety	3.7
Schools	3.1	Safety	3.4	Air	3.4
Air	2.9	Space	2.7	People	3.2
Space	2.7	People	2.6	*Shopping*	2.8
People	2.4	*Shopping*	2.5	Space	2.8
Shopping	2.2	*Work*	2.2	*Friends*	2.6
Work	1.9	*Friends*	1.8	Density	1.7
Friends	1.8	*Recreation*	1.8	Schools	1.6
Recreation	1.8	*Culture*	1.7	*Culture*	1.3
Culture	1.5	Density	1.5	*Recreation*	1.3
Density	1.5	*Schools*	1.1	*Work*	0.6

[a]Adjusted for the population groups (see explanation in Table 9). Italicized attributes are access-related. (Values in each column add up to 25.)

SUMMARY AND CONCLUSIONS

In this chapter, we have presented the common themes in public perceptions of the residential area. Our open-ended questions permitted us to elicit what people valued, rejected, or desired in their environments without prejudging the relevant issues or the basis of what planners and sociologists have deemed important. Moreover, these questions provided a schema for thinking about the primary organizing elements necessary for alternative design concepts. The more structured questions permitted a preliminary inventory and comparison of residential quality, providing important background information for understanding satisfaction, preferences, and priorities. They also established a baseline measure for a comparison of the "holdings" of various groups, a baseline that is necessary for understanding a sense of the deficits, surpluses, and inequities between and among the various groups.

One of our concerns was to identify the underlying organizing principles in people's minds about what constitutes residential existence and well-being. Mindful that in the past the "experts" or professional environmental designers might have prejudged the issue, thereby unwittingly foreclosing consideration of some important facets of residential living, we sought to elicit these principles to guide our own subsequent formulation. Recalling the common themes that we uncovered (social milieu, crime and safety, air quality, amenities, conveniences, appearance, atmosphere, and services) we found some discrepancies with the premises of the original neighborhood-unit formulation.

We saw, for instance, that just as the sociologists have maintained, the residential area is primarily conceived of as a social milieu, with physical design being secondary. Thus, one's neighbors are of more concern than the layout of the area or the facilities that it contains. This finding will be an important consideration when we discuss alternative organizing concepts for residential areas, for it raises the issue of the interrelationships between social policies and environmental design. Although these have not been explicitly linked in the past, even where acknowledged, new design concepts must join them, even if the nexus is implemented only through physical design.

In a related vein, we found that crime and safety issues are also important concerns in residential life. Although these were not a concern when the neighborhood unit concept was formulated, they have been on everyone's mind in the latter half of the twentieth century. But the specifics of this concern were related to the type of people who live in an area, not the physical design that might facilitate or impede antisocial behavior or criminal acts against people and property (Newman, 1972).

Although the major concern with air quality expressed by our respondents may have reflected a particular problem of the Los Angeles region, it is similar to the first two themes in that solutions to these problems lie beyond the control of a design concept that focuses only on portions of a larger urban area. Social patterns of settlement, as well as social behavior, are caused by forces largely beyond the control of the traditional neighborhood designers. They are metropolitan and national in origin and depend on citywide and nationwide policies for their control. Air quality is largely a problem that requires regional and national policies rather then residential area ones. Thus, any new design concept must be more cognizant of the importance of these larger concerns.

The next four themes—amenities, conveniences, appearance, and atmosphere—are precisely the concerns embodied in the neighborhood unit. Here, we found people expressing themselves in pretty much the way that environmental designers have imagined when they have provided residential design solutions. Our findings, in the main, pertain to refinements and adjustments in these traditional areas.

The last item, the residential area as a conduit for services, looks at a different aspect of providing for residential well-being. Although the neighborhood unit concept, as well as any other design concept, is largely oriented toward providing capital improvements, the service aspect pertains to operating budgets and levels of maintenance and service once capital improvements have been put in place. Operating and maintenance concerns suggest that any design concept must include considerations of its long-term functioning without requiring unnecessarily high operating costs. In short, design concepts must be formulated not only to achieve a high initial positive impact but also to continue to produce that impact, at a reasonable cost, over the life of the product.

We might add parenthetically that it was surprising to us that two themes in particular did not surface in these interviews to any significant extent. Given the booming Southern California real-estate market of the middle to late 1970s, we expected the investment aspects of the residential environment to receive considerable emphasis, at least from the upper- and middle-income respondents, but no such mention was made in our interviews. Yet, according to a 1976 study of neighborhood environments conducted in the metropolitan areas of Houston, Dayton, and Rochester, this concern was very much in the mind of upper-status people, who seemed to be seeking a neighborhood where "you can anticipate resale at an appreciated value" (Coleman, 1978, p. 10). Is it possible that this concern was not uppermost in people's minds in the early to middle 1970s in Southern California? Or were investment values indirectly reflected in the themes made more explicit? Similarly, the quality of public schools was not emphasized as such in our study, despite the smoldering school-desegregation controversy of this period, whereas in the Coleman study, both good public schools and proximity to the community's prestigious private schools were of concern to the upper-status people. According to this study, the middle-income people also emphasized quality of public schools. Yet, this point was not brought up in our interviews in any noticeable way. This is indeed a baffling aspect of our findings.

With the implications for these underlying residential themes in mind, we can turn to reviewing the expressed preferences of the different groups and their baseline holdings. We saw, for instance, the smug contentment of the upper-income group, who could afford to live in settings that satisfied basic residential-quality requirements: a preferred social context, good air, comparative personal and property safety, and a good physical place that abounded in amenities and services. We observed the uncertain satisfaction of the middle-income groups, who resided in areas that were less than optimal but were nevertheless satisfactory on most counts. And we learned of the fears and anxieties of the lower-income groups, who, by comparison, had the worst residential areas, frequently unsatisfactory according to most criteria of residential choice. These impressions, elicited in the course of the respondents' describing their areas, also correspond closely to their overall *evaluations* of their areas on a five-point scale, as shown in Table 3.10. This table highlights the findings presented in the previous tables and in the text, and it points out the shortcomings of the respective residential environments as felt by the different groups. As we have noted previously, the differences among the various racial groups, once income was controlled, were usually not significant. The low-income blacks, however, appeared to live under the worst environmental conditions. It seems likely that although the range of residential choice was quite limited among the poor in general, for poor blacks it was simply nonexistent. Furthermore, we find that within a particular income and ethnic group the differences in baseline qualities available to different stages

TABLE 3.10. Overall Evaluation of Residential Areas by Population Groups

	Percentage of respondents who considered their residential area					
	Excellent	Good	Average	Fair	Poor	n
Upper-income white	85.9	11.8	2.4	—	—	85
Middle-income white	32.5	46.3	16.3	1.3	3.8	80
Middle-income Hispanic	15.3	55.9	25.4	3.4	—	59
Middle-income black	12.8	46.5	27.9	12.8	—	86
Lower-income white	4.5	21.6	44.3	17.0	12.5	88
Lower-income Hispanic	14.5	29.1	32.7	16.4	7.3	55
Lower-income black	4.5	9.1	4.5	40.9	40.9	22

in family cycle were small; the differences that existed can be attributed to variable expectations and standards of evaluation resulting from lifestyle differences.

The baseline inventory of residential qualities using separate indicators also suggests that the concerns and priorities of different population groups were somewhat different (Table 3.11). Thus, it can help to define the appropriate issues for the residential areas of different population groups.

It further helps to identify those issues that require public policy responses

TABLE 3.11. Aspects of Residential Quality with Which a Majority of Respondents Were Less than Satisfied (Ranked According to Frequency of Mention in Each Group)

Upper white (n = 85)	Middle white (n = 80)	Middle Hispanic (n = 59)	Middle black (n = 86)	Lower white (n = 88)	Lower Hispanic (n = 55)	Lower black (n = 22)
None	Smoggy	Smoggy	Smoggy	Crowded	Smoggy	Unsafe
	Crowded	Crowded	Crowded	Smoggy	Crowded	Smoggy
	Noisy	Noisy		Lack of privacy	Lack of privacy	Lack of privacy
				Ugly	Ugly	Dirty
				Neglected	Impersonal	Noisy
				Dirty	Undesirable people	Ugly
				Noisy		Impersonal
				Impersonal		Crowded
				Undesirable people		Undesirable people
				Unsafe		Inconvenient
						Lack of dwelling space

at the city or metropolitan level. For example, problems of noise, privacy, and density can be mitigated by specific design responses at the local level. Such problems as smog, safety, and maintenance services can have local physical design responses but must be supplemented by citywide strategies extending beyond the boundaries of any one residential area. On the other hand, problems of social relations and milieu usually lie completely beyond the scope of local physical planning and can be addressed only by means of city or metropolitan policies of blending social groups through redevelopment, neighborhood conservation, and inclusionary zoning requirements.

Moreover, the baseline of residential amenities possessed by each group helps us to appreciate the differences in residential perceptions presented in subsequent chapters. An understanding of these differences is important in thinking about appropriate policy responses to inequalities in residential holdings. For instance, although the old *Planning the Neighborhood* standards and requirements were geared to an absolute notion of residential quality, we will see that many of our respondents thought in terms of marginal improvements of what they already possessed. Furthermore, these differences in current holdings suggested differences between two strategies for the design of future development: those of a compensatory nature that would provide countervailing improvements in those residential settings currently deprived on a comparative basis; or those that, in response to consumer choice rather than income or social inequalities, suggest policy and design measures to counterpoise attributes existing in current residential settings. In short, these findings suggest the whole policy issue of unequal responses directed at different groups to correct or mitigate current inequalities.

We return to these policy and design issues at the end of the book, but first, we must consider other aspects of residential perceptions and preferences. In the next chapter, we turn to a different dimension of presenting perceptions of and preferences for residential amenities. Here, we examine how people image their residential areas and the distinguishing elements of the collective images of different localities.

NOTES

1. The three groups comprised approximately 90% of the total population of the region.
2. The stage in the family cycle was controlled by screening interviewees during the initial stage. As it was generally difficult to find families with older children, the initial sampling scheme, which called for families with young children as one group and families with older children as another, was revised so that these two groups were included in the single category shown here.
3. Some 244 interviews were completed by May of 1972, at which time the Public Health Service, which had funded this first stage, had its appropriations for this kind of research eliminated.

With subsequent funding from the NIMH, we completed the additional 231 interviews in the spring of 1974. We also undertook the interviews of the low-income Hispanic and the low-income black groups during this second stage of interviewing. See Table A2, Appendix II.

4. We used an abbreviated interview schedule (see Appendix I) for these two low-income groups because of the difficulty in getting members of these groups to agree to an interview at all, despite our payment of ten dollars for their time (none of the material was eliminated for the questions we report on here). Because of our difficulty in finding elderly low-income Hispanic respondents, some of the interviews were conducted in the Maravilla Public Housing Project.
5. The areas were chosen after we first determined tentative locations from a review of preliminary 1970 Census returns, assessors' maps (housing value was used as an indication of income), and discussions with people knowledgeable about the metropolitan area. Following the preliminary identification, field checks were made to screen out areas not basically residential.
6. Lists of addresses were obtained from the Public Systems Research Institute of the School of Public Administration, University of Southern California.
7. About 30% of the households initially contacted were acceptable and available for further inquiry. Of these 30%, 65% agreed to participate. These were the major problems encountered: (1) the "wrong" race was frequently found in supposedly middle-income Hispanic areas; (2) no one was at home during the day in middle-income white areas; and (3) the potential respondent in the lower white and middle black areas refused to participate. Other common problems included "income too high" and "refusal during initial contact."
8. We were advised that the very length of our survey instrument and the degree of cooperation and interest required of the respondents in order to complete the interview meant that our final set of interviews would not be strictly randomly selected, no matter what our sampling scheme. Dr. Virginia A. Clark, Professor of Bio-Statistics of the School of Public Health, University of California, Los Angeles, provided valuable assistance in helping us formulate our sampling design.
9. See Appendix I. For parsimony in data, we have collapsed the seven-point scales into a dichotomous form whereby, for example, "extremely desirable," "desirable," and "somewhat desirable" are categorized as one class, and "neither desirable nor undesirable," "somewhat undesirable," "undesirable," and extremely undesirable" are categorized as the other. Thus, a neutral response was assumed to be the threshold to cross in order to indicate at least some positive response. This premise is applicable to all other data display formats in this chapter and elsewhere in this book where similar measures have been used.
10. According to the 1970 Census data, which were the basis for the selection of our interview areas, all but the middle-income black and Hispanic areas were considered fairly "homogeneous" in terms of racial mix. By the time we were conducting the second phase of our interviews in 1974–1975, it was apparent from field visits that some of the low-income white areas were being inhabited by the Hispanic population. Thus, although these responses may reflect some exaggerations of racial mix, they may not be entirely inaccurate.
11. From a two-way analysis of variance, where both factors, "population type" and "stage in family cycle," were introduced simultaneously.
12. This attitude was apparent in the difference in the adjusted means for the three stages in the family cycle for these two scales. Although the stages in the family cycle were not a significant factor in the case of the "poor-rich" and the "high-status–low-status" scales, the two-way interaction between population type and stages in the family cycle was statistically significant. A breakdown of means by population type and stages in the family cycle showed a consistent pattern of interaction for both of these scales. In both cases, the elderly of the upper- and middle-income groups showed a more negative evaluation than did the rest, and the elderly of the lower-income groups showed a slightly more positive evaluation than their low-income counterparts. That is, the interaction pattern, schematically, was as follows:

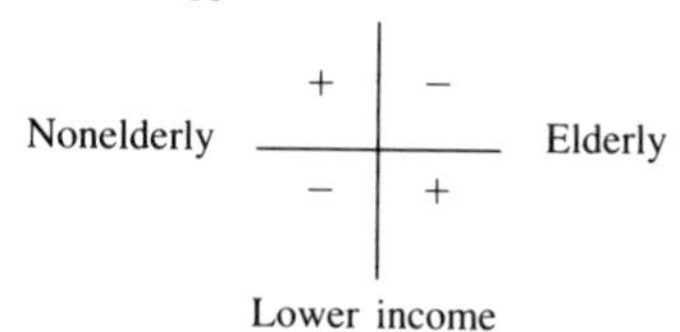

13. However, there was a significant interaction effect between population type and stages in the family cycle. A breakdown of means (using the original uncollapsed seven-point scale) by population type and stage in the family cycle showed an interaction pattern similar to the ones reported in note 12. That is, the elderly subgroup of the middle- and upper-income groups tended to evaluate their areas more negatively than respondents in the other two stages in the family cycle in the same income class. Their lower-income counterparts, however, tended to evaluate their environment more positively than the other low-income family-cycle subgroups. Perhaps the upper- and middle-income elderly, although considering their area safer than the low-income areas in general, considered themselves more vulnerable, whereas their low-income counterparts—because they had so little in terms of material possessions—felt less vulnerable to crime.
14. Compare with note 13.
15. In addition, interaction effects were found to be significant in the case of the ''comfortable-uncomfortable,'' the ''beautiful-ugly,'' and the ''neglected-cared for'' scales. A breakdown of unadjusted means by population type and stages in the family cycle suggested a very similar relationship between population type and stages in the family cycle, as noted in note 13, although the patterns were less sharp than before.
16. See Question 13—the ''trade-off game''—of the questionnaire shown in Appendix I.
17. Also from the same section of the questionnaire (see note 13).

4

Residential Area and Neighborhood

Images and Values

The previous chapter focused on verbalized concepts and evaluations of, and preferences for the residential environment. Although verbal methods of expression lend themselves to many aspects of description and evaluation, they are less effective in describing the relationships of objects in space. Yet, such relationships are the core of any environmental design construct. To tap these aspects of the residential environment, we focus in this chapter on the physical-spatial characteristics of the residential settings by examining people's image of their residential area, as portrayed on the maps that we asked them to draw.

These cognitive maps served several purposes. They allowed us to understand further the basic constructs that people used to organize their residential area conceptually. Just as the neighborhood unit was a concept that required both words and maps to explain, we also needed both modes of representation to describe and depict underlying organizing constructs. Additionally, the cognitive maps facilitated exploring relationships between the "residential area" and the "neighborhood," where these terms evoked different constructs in people's minds. Finally, these cognitive maps provided a context for examining residents' views on the importance of living in a "neighborhood." In short, these perceptions provided necessary supplements to the themes explored in the previous chapter.

Like their verbal responses, the maps drawn by the residents were quite diverse and reflected many different styles of "structuring" (Appleyard, 1969, 1976) and conceptualization. Some of these maps were schematic and abstract, others were detailed and pictorial. Although some maps showed accurate cartographic representations of the street network, others lacked accuracy and scale relationships, emphasizing only those places and points in space that were personally relevant to the respondent.

These maps also varied in content and extent. To give some examples, a Venice resident drew only a street intersection (Figure 4.1); a City Terrace resident drew only a street block (Figure 4.2); a Palos Verdes resident showed the whole peninsula (Figure 4.3); and a Bel-Air resident drew an entire section of

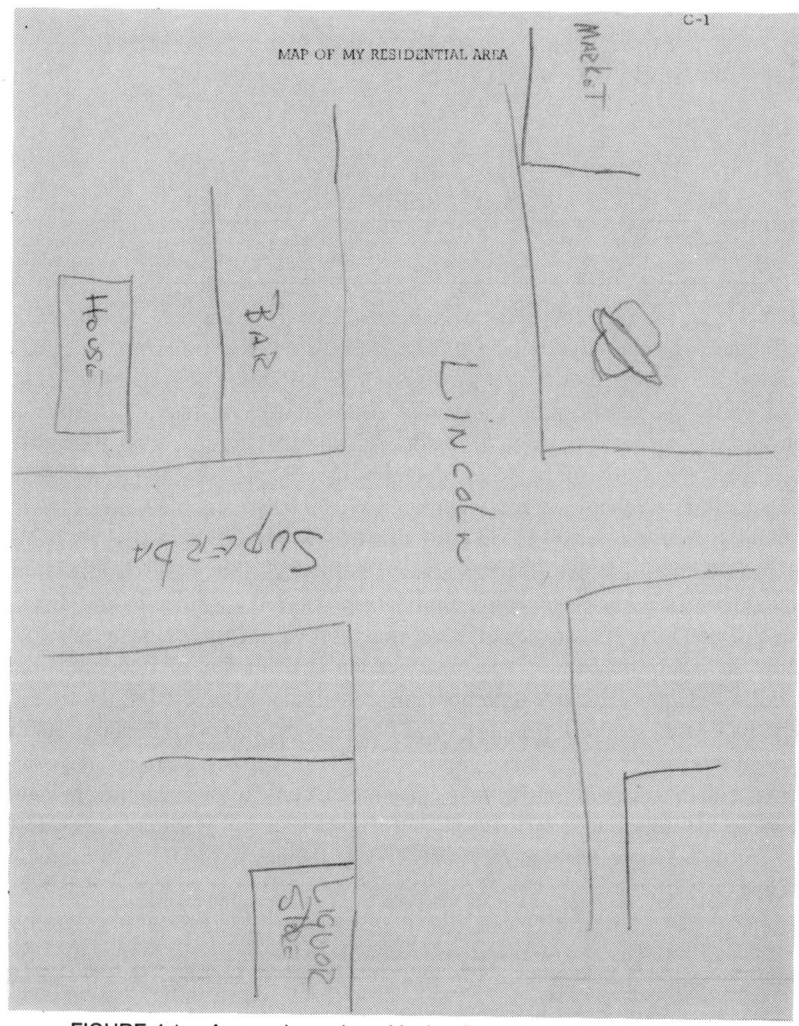

FIGURE 4.1. A map drawn by a Venice (lower-income white) resident.

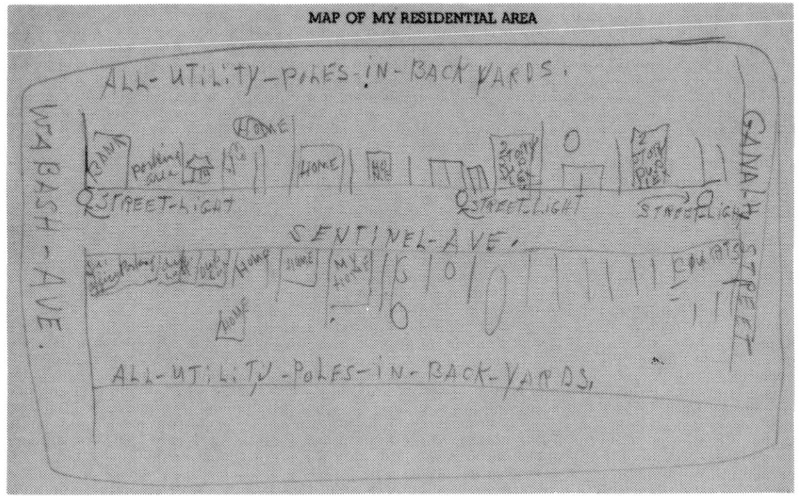

FIGURE 4.2. A map drawn by a City Terrace (lower-income Hispanic) resident.

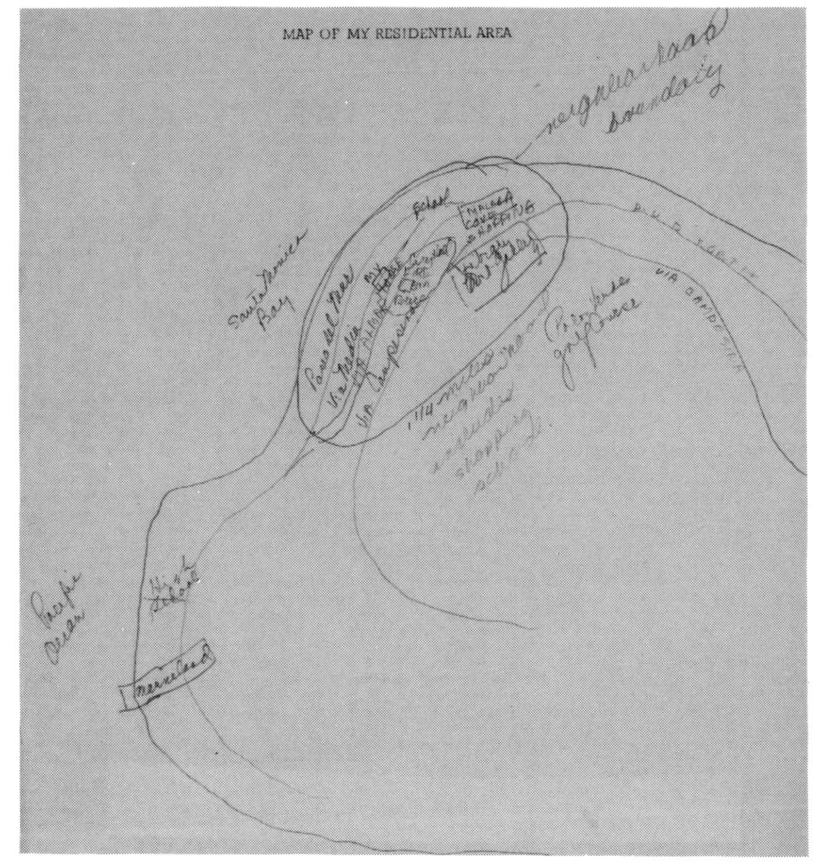

FIGURE 4.3. A map drawn by a Palos Verdes (upper-income white) resident.

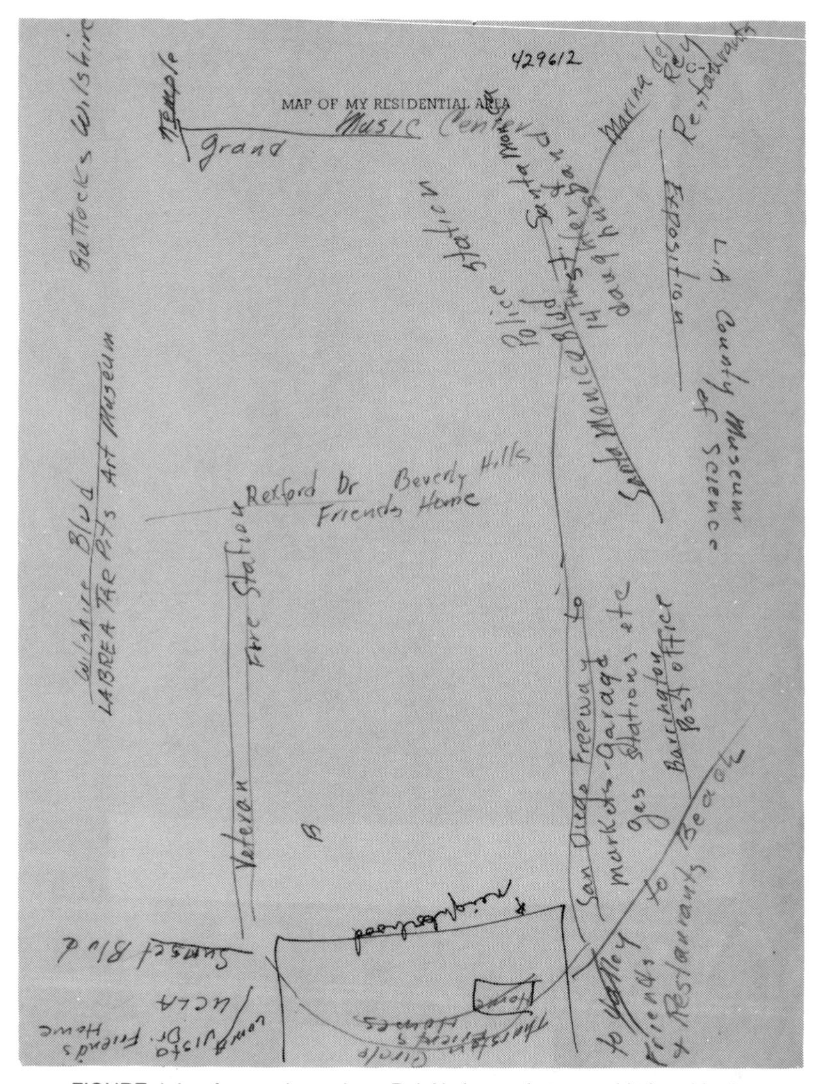

FIGURE 4.4. A map drawn by a Bel-Air (upper-income white) resident.

the city stretching from the west side to downtown Los Angeles—an area large enough to constitute a medium-sized city (Figure 4.4).

Why such disparities? Why such wide differences in detail, pictorialization, style, and content? Many of these maps appeared to illustrate different theoretical formulations about social and psychological constructs of the residential area.

Consider, for example, the map drawn by the Venice resident (Figure 4.1).

It shows the intersection of Lincoln Boulevard, a major arterial in West Los Angeles, and Superba Avenue, a local cross street. All this person acknowledged as his residential area was a street intersection in a large metropolis. Why? Several explanations are possible, based on the available theoretical views. Some authors (Horton and Reynolds, 1971; Buttimer, 1972) have argued that the cognitive map of the relevant physical environment is shaped by the space to which our actions and behaviors are limited. According to this view, the intersection of Lincoln and Superba may indicate a severely limited "action space,"[1] suggesting a physical handicap or an age-related incapacity in the person who drew this map. Another possibility is that the person may truly have belonged to a "nonplace" realm as discussed by Webber (1964).[2] If so, the immediate place had little significance to this person: all he needed to identify where he lived was the nearest street intersection, a mere reference coordinate in a "nonplace" realm rather than a symbol of a residential place.

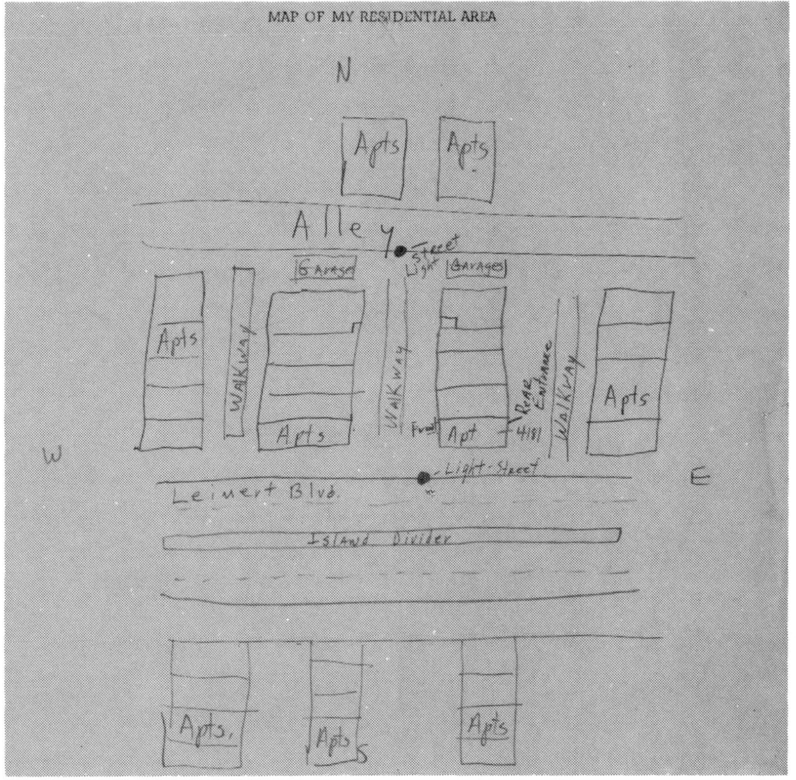

FIGURE 4.5. A map showing an apartment complex as the residential area—drawn by a Crenshaw (middle-income black) resident.

The person who draws a street block (or a collection of street blocks) as his residential area may be conveying still another view (e.g., Figure 4.2). Here, the image of the residential area is a reflection of the "face-block" community, which Suttles (1973) discussed in detail. The sense of a "face-block" community is derived from familarity, nodding acquaintances, and neighborly interactions. It is likely that here the image of the place is shaped by this sense of community. The residential area as reflected in these maps is a space defined by propinquity and social encounters.

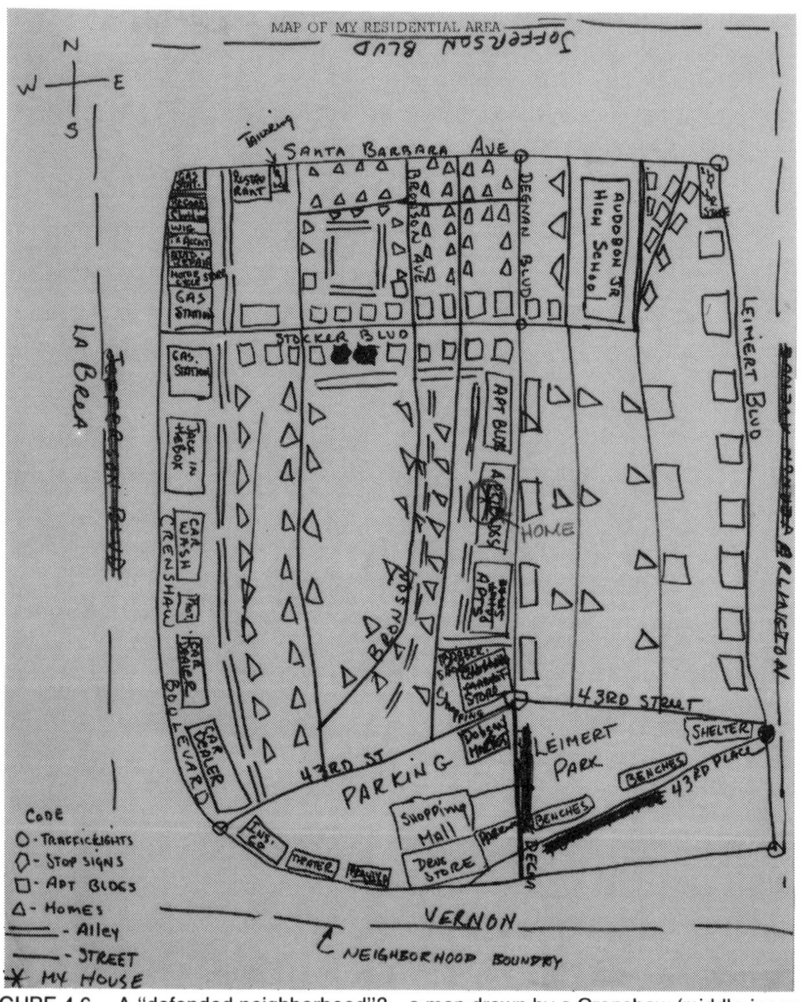

FIGURE 4.6. A "defended neighborhood"?—a map drawn by a Crenshaw (middle-income black) resident.

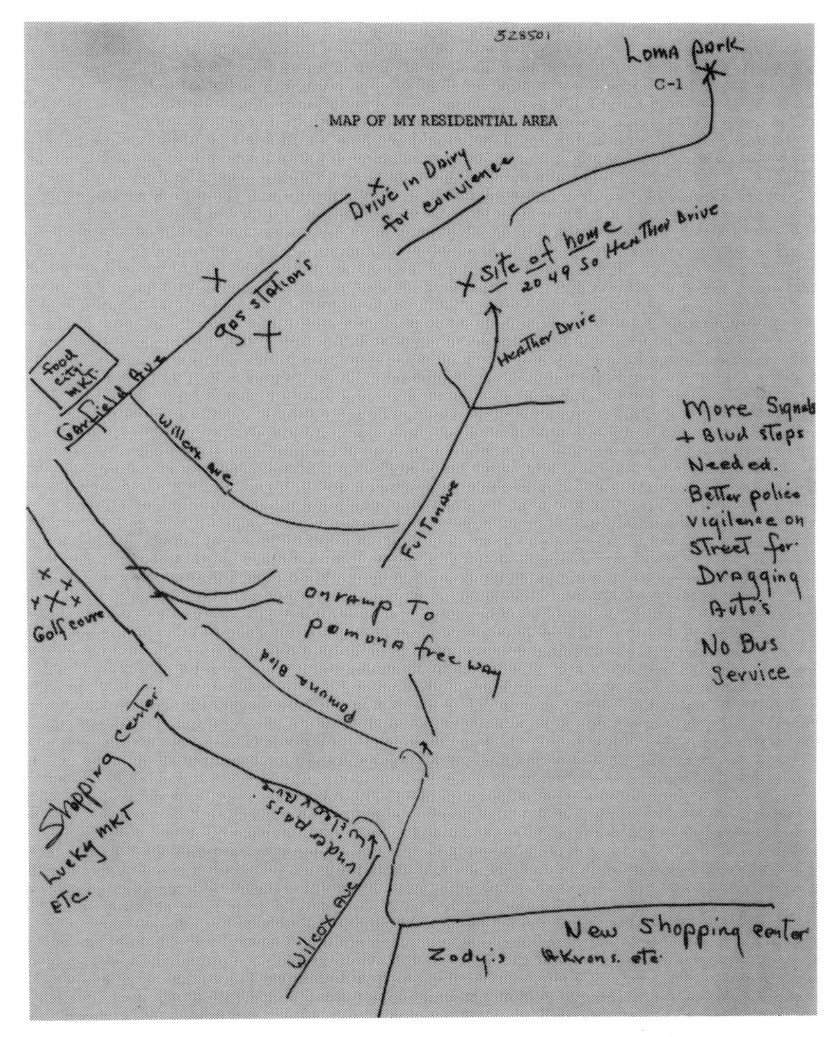

FIGURE 4.7. Cognitive map of a Monterey Park (middle-income Hispanic) resident.

The person who shows the apartment complex as his residential area (Figure 4.5) or the person who draws a residential area map with a rigidly drawn boundary (Figure 4.6) may be conveying a sense of a "defended neighborhood," to use the terminology of Suttles (1973) again. The image of a defended neighborhood is of a territorially defined, well-framed, well-bounded place. The thresholds are all-important in this schema. They are sharp and distinct, separating one's residential area from the rest. There is little ambiguity about where the residential area begins and where it ends.

Then, we find maps that are simply a loose collection of points, places, and landmarks, strung together by a partially drawn street network (Figure 4.7). These maps appear to mirror the residential "behavior circuits" (Barker, 1968; Perin, 1970)—the locus of daily activities and their settings, which become one's immediate residential domain.

We also find maps that are nothing more than cartographic representations of the locality where the respondents lived (Figure 4.8). These maps typically show the street network with an occasional landmark highlighted here and there.

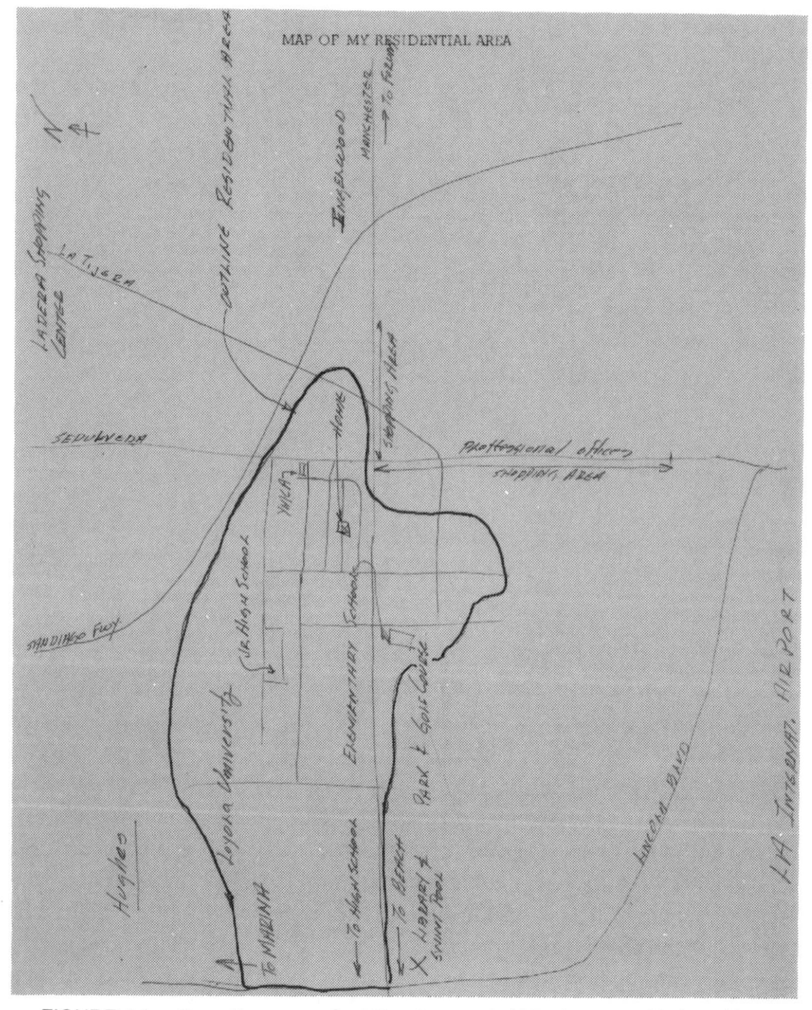

FIGURE 4.8. Cognitive map of a Westchester (middle-income white) resident.

Usually, these maps have no edges, no focal points. They reveal a definite knowledge of the street network but appear indefinite about where the residential area begins or ends. They may at best convey a notion of "home area,"[3] but nothing more. These maps suggest neither a territorial identity nor a sense of "turf." It is possible that these respondents also belong to a "nonplace" realm, lacking any real sense of attachment to the immediate residential setting. Alternatively, these maps may reflect the developing image of a stranger who has learned the spatial structure of the environment but has yet to ascribe her or his personal meanings and values to the place (Lynch, 1960).

This typology of residential area maps is not definitive because the categories are not mutually exclusive. A large number of maps display characteristics of more than one of these types; others simply defy categorization. We offer the typology as a guide to thinking about them and the meanings that they may portray.

To summarize, we found that many different spatial constructs were presented when we asked people to draw maps of their residential areas. In some cases, the area was depicted by a mere intersection, in others by a city-block face, a whole block, or a series of blocks. In some instances, the territory was well demarcated and well defined; in others, it was shown as a loose collection of points, places, and landmarks strung together by a partially drawn street system with little sense of where the area began or ended.

Rarely was the area portrayed in the systematic and hierarchical fashion embodied in the neighborhood unit, with its clear demarcation of boundaries and an inward focus around some kind of central core of public facilities. The salient features found in the designers' highly rationalized construct are only partially realized in the minds of most people.

COLLECTIVE IMAGES

In turning to the "collective images"[4] of the twenty-two settings, we find the inescapable signature of the automobile, which has influenced so much of the Los Angeles metropolitan form. Streets, highways, and freeways were the dominant organizing elements in these collective images. Of these, the arterial streets were most frequently mentioned. Shaped by the Jeffersonian grid system and built to overly efficient traffic-engineering standards, these seemingly endless corridors lace the metropolitan area in quarter- or half-mile intervals. Many of them have developed as "strip commercial," and although they are not unique to the Los Angeles area, their dominance here has been an object of derision and contempt among the critics of urban aesthetics. Nevertheless, they seem to have played an important role in the collective cognitive maps of our Los Angeles respondents. In a vast, sprawling urban space, these arterials are major frames of reference, important anchors of orientation in space.

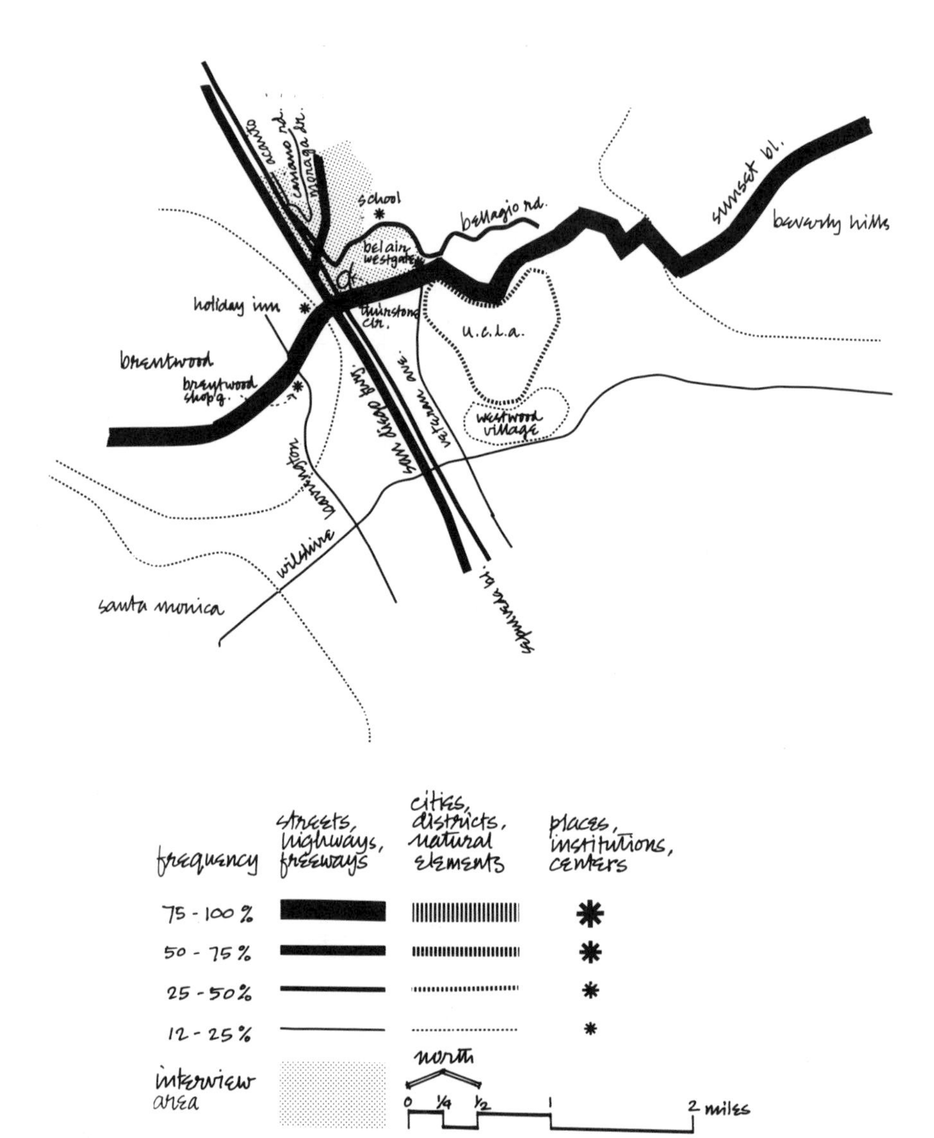

FIGURE 4.9. Collective image—Bel-Air (upper-income white). NOTE: The legend also pertains to Figures 4.9–4.30.

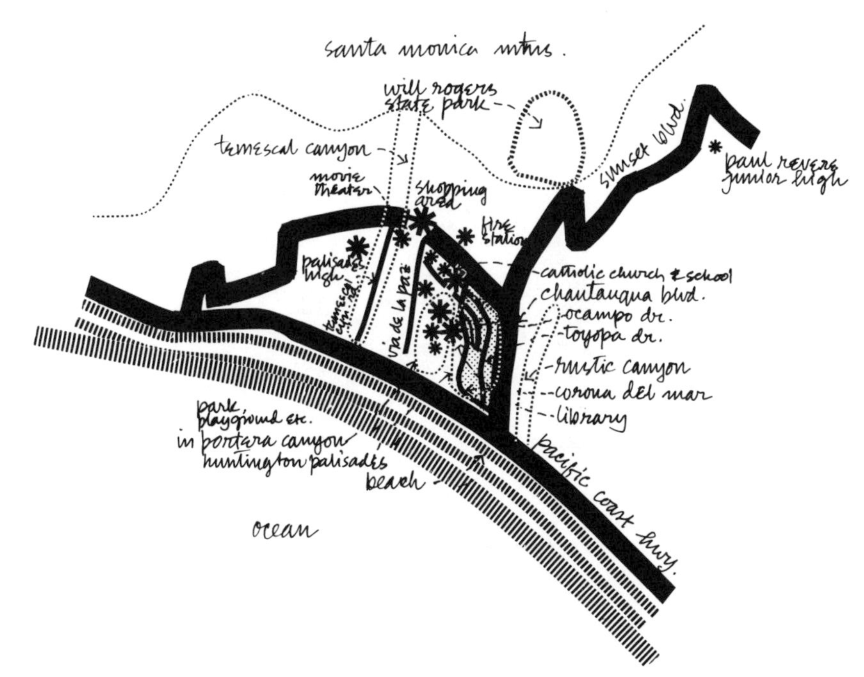

FIGURE 4.10. Collective image—Pacific Palisades (upper-income white).

Of course, the ubiquitous street system was organized in different ways in the minds of our respondents, but these divergences appeared to originate in differences in the physical layout of the various areas rather than in distinctions between the groups by income, race/ethnicity, or stage in the family cycle.

Thus, in some of the collective maps, we found that one or two arterials became the all-important axes or "spines" of the residential area image. Because these were important routes of travel, they also became the conduits of commercial and institutional land uses and services, thereby becoming an important activity spine. In these collective images, streets had a definite hierarchy of importance, and the residential area was essentially represented as a hierarchically organized street system. The collective images of Bel-Air (Figure 4.9), San Marino (Figure 4.12). Montebello (Figure 4.19). Whittier (Figure 4.21), Westchester (Figure 4.18), and Watts (Figure 4.23) reflect this hierarchy dominated by a single arterial.

In other cases, no single street dominated, and a number of major routes of travel were mentioned with equal frequency. The residential area image thus became a "net" or "web" of streets and highways in a nonhierarchically orga-

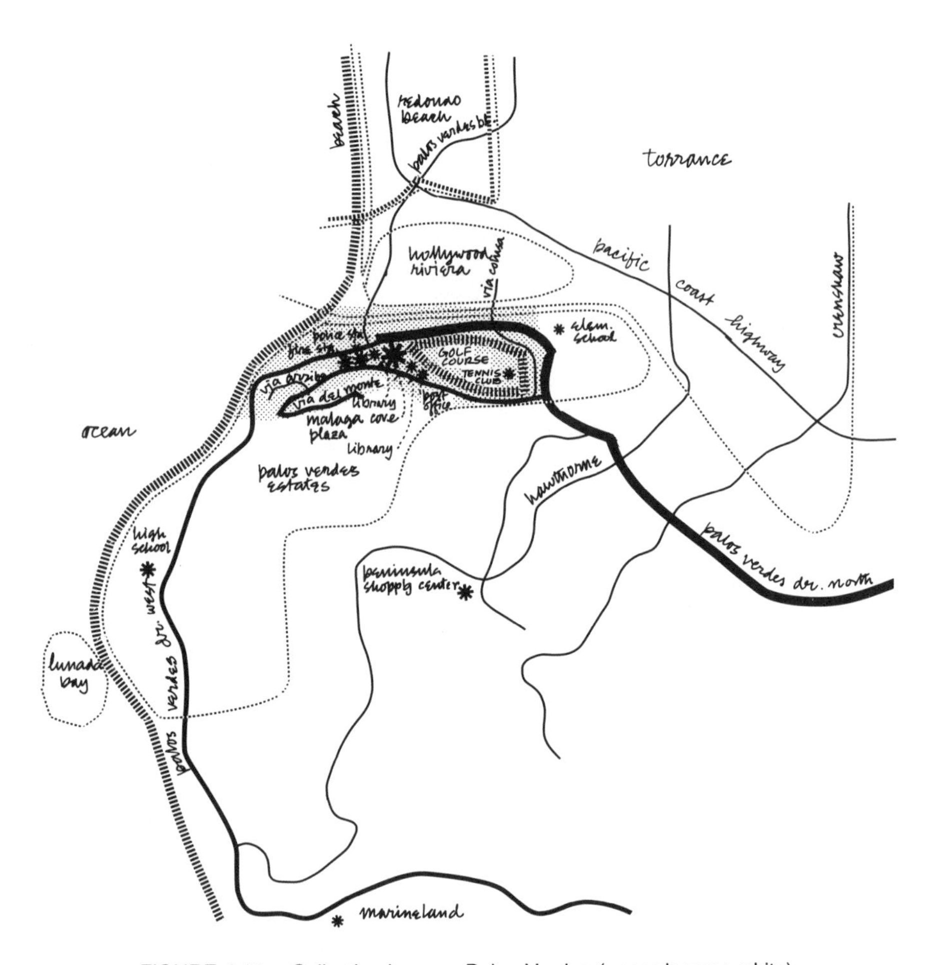

FIGURE 4.11. Collective image—Palos Verdes (upper-income white).

nized conception of the residential area. The collective images of Van Nuys (Figure 4.17), Boyle Heights (Figure 4.28), Temple City (Figure 4.16), and Baldwin Park (Figure 4.24) are some key examples.

None of these collective maps offers a strong sense of boundary definition. Only in settings located near the coast or the mountains did elements of the natural landscape become obvious edges. The coast was a strong edge in the collective images of Palos Verdes (Figure 4.11) and Pacific Palisades (Figure 4.10) residents. The Monterey Hills appeared as a minor edge in the collective image of the Monterey Park residents (Figure 4.20), and so did the Santa Monica Mountains in the images of the Pacific Palisades (Figure 4.10) and Bel-Air

(Figure 4.9) residents. The San Gabriel River—thoroughly paved for precautionary flood control measures—played a similar role in the collective image of the East Long Beach residents (Figure 4.15).

Very few of these collective images project a sense of a "core" or a center. There is a hint of such a focal point in the maps of Palos Verdes (Figure 4.11) and Pacific Palisades (Figure 4.10) and, to a lesser extent, in the maps of Westchester (Figure 4.18) and San Marino (Figure 4.12). Typically, this focal organization was made up of a constellation of important public buildings, spaces, and commercial establishments in proximity to each other.

Minor nodes were present in some other maps, but usually these were intersections of main arterials. A typical minor node, for example, consisted of four gas stations in the four quadrants of an intersection, often accompanied by a cluster of supermarkets, fast-food stands, drugstores, Laundromats, liquor stores, and the like. These nodes were most commonly found in the maps of Van Nuys (Figure 4.17), Temple City (Figure 4.16), and Westchester (Figure 4.18). Elsewhere, community business districts served as major nodes, and they included other forms of civic and public facilities as well. Such elements played a

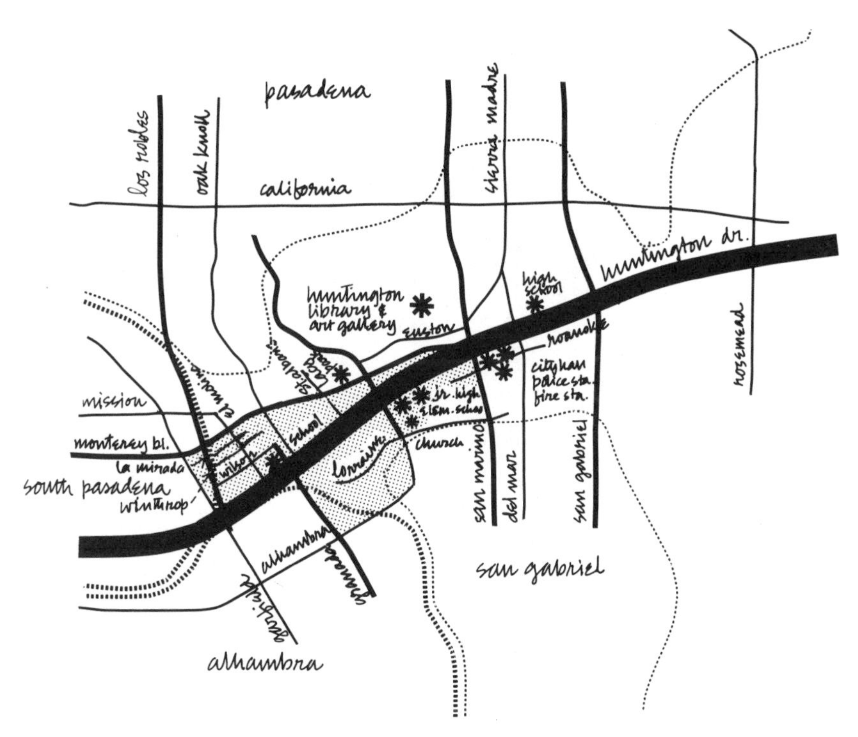

FIGURE 4.12. Collective image—San Marino (upper-income white).

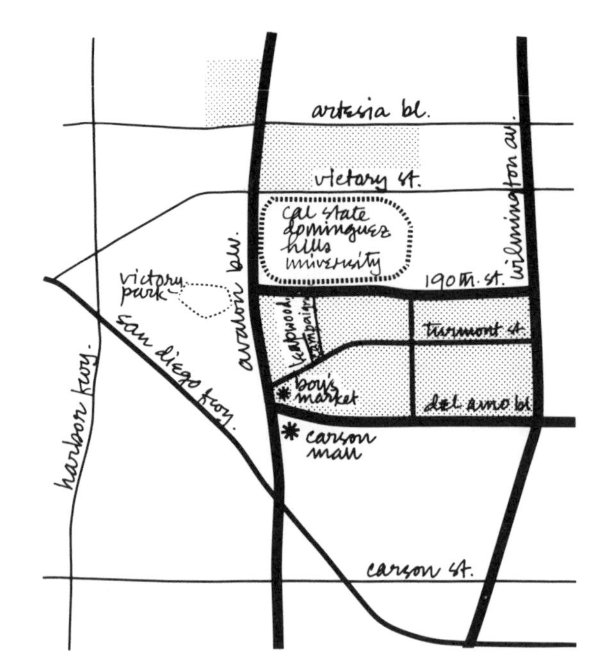

FIGURE 4.13. Collective image—Carson (middle-income black)

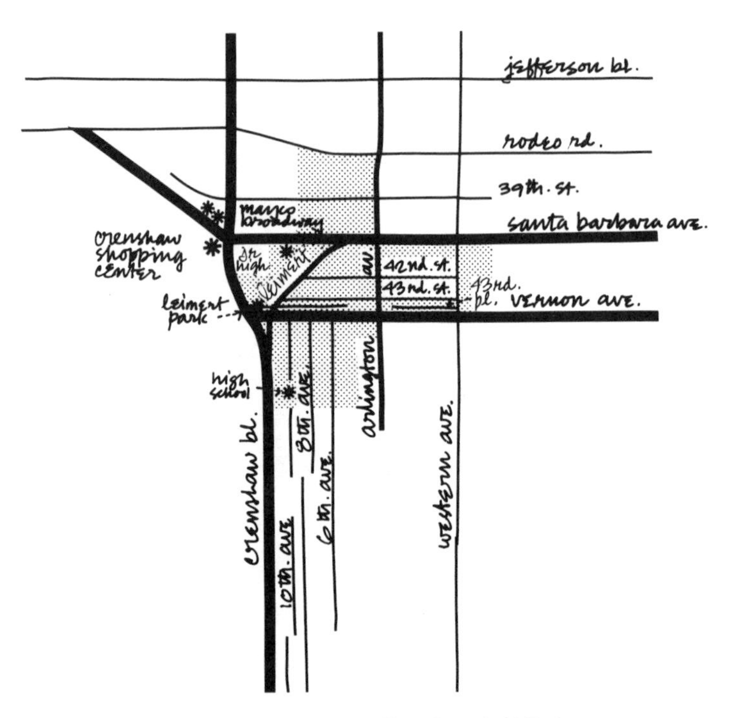

FIGURE 4.14. Collective image—Crenshaw (middle-income black).

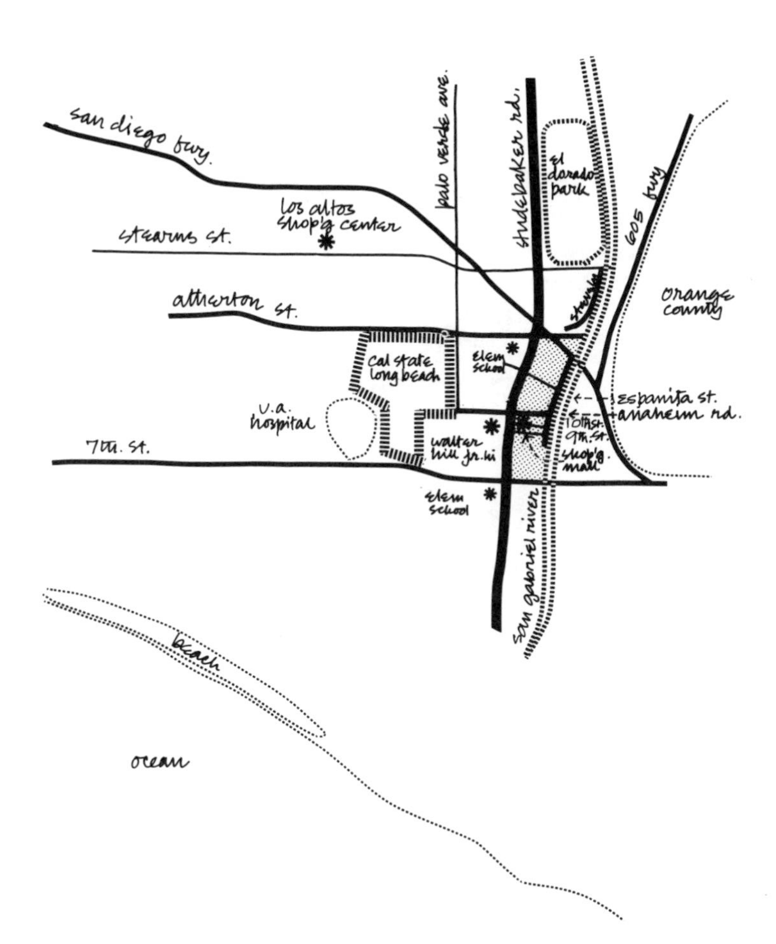

FIGURE 4.15. Collective image—East Long Beach (middle-income white).

dominant role in the collective images of Pacific Palisades (Figure 4.10), San Marino (Figure 4.12), Palos Verdes (Figure 4.11), Westchester (Figure 4.18), Crenshaw (Figure 4.14), and Temple City (Figure 4.16). In other settings, large regional shopping centers appeared as major nodes of activities. Thus, the Peninsula shopping center in Palos Verdes (Figure 4.11), the Los Altos shopping center in East Long Beach (Figure 4.15), Carson Mall in Carson (Figure 4.13),

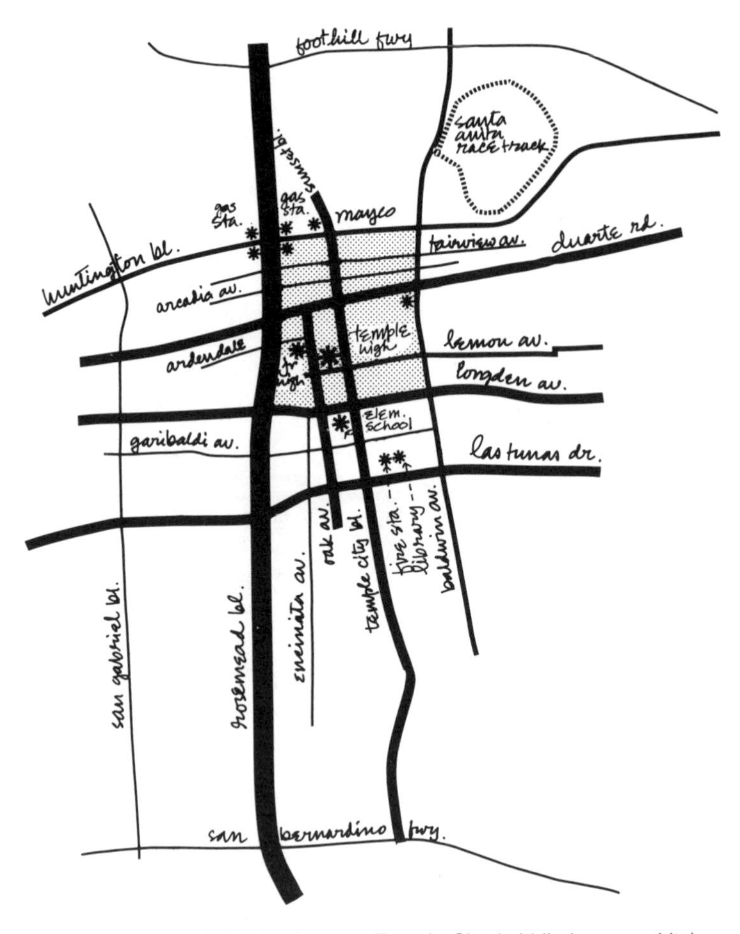

FIGURE 4.16. Collective image—Temple City (middle-income white).

the Crenshaw shopping center in Crenshaw (Figure 4.14), Montebello Mart in Montebello (Figure 4.19), and the Atlantic Square shopping center in Monterey Park (Figure 4.20), all served as major nodes.

These shopping centers, because of their sheer size, were also important landmarks in the urban landscape. In terms of visual significance, they seemed to outclass such public facilities as schools, fire stations, libraries, police stations, and city halls. Nevertheless, these public facilities played a role as minor landmarks. Parks and playgrounds, where present, played a similar role. Occasionally, large public institutions, such as community colleges, universities, or hospitals, appeared as important reference points in the collective images of the residential area.

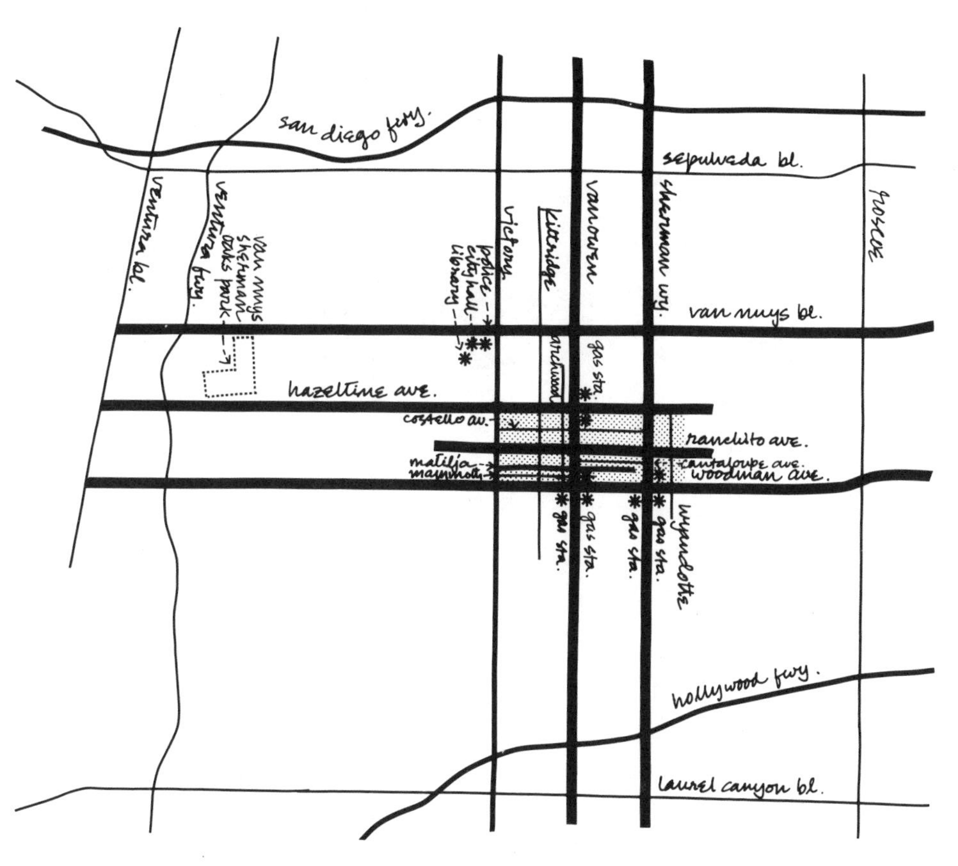

FIGURE 4.17. Collective image—Van Nuys (middle-income white).

Some of these maps also contained occasional reference to such identifiable and well-known "districts" as Westwood Village (in the Bel-Air map, Figure 4.9), Huntington Palisades (in the Pacific Palisades map, Figure 4.10), Hollywood Riviera (in the Palos Verdes map, Figure 4.11), and Marina del Rey or the Los Angeles International Airport (in the Westchester map, Figure 4.18). The adjoining political jurisdiction often provided an identifiable, if symbolic, frame of references in these maps. Thus, such referents as Orange County, Beverly Hills, and San Gabriel became important components of the collective images of the residential area.

An examination of these area images offers some clues about how the identity of a residential area may be derived. But the stress placed on the street system and the nodes and the remarkable lack of salience of boundaries leave us

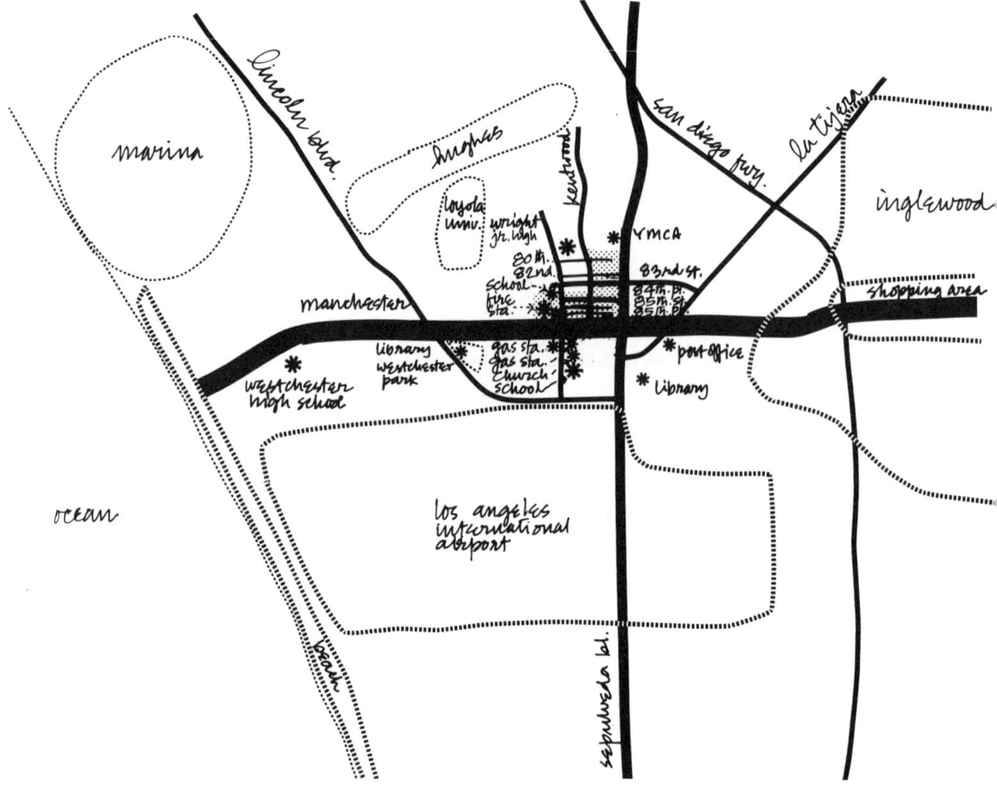

FIGURE 4.18. Collective image—Westchester (middle-income white).

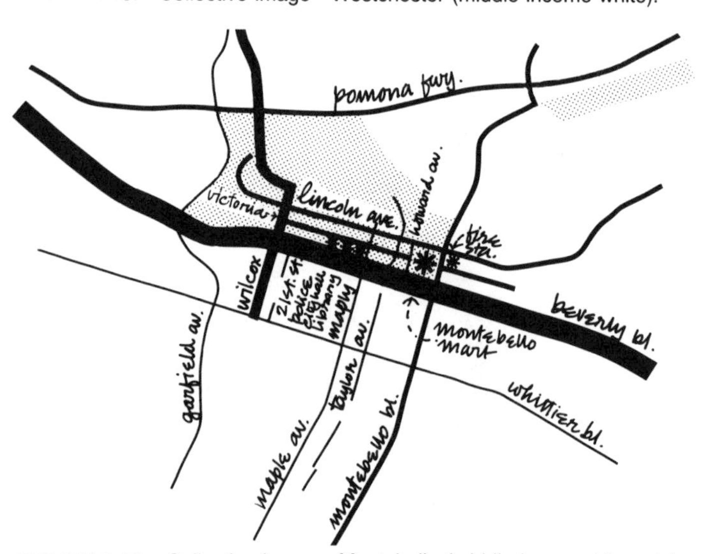

FIGURE 4.19. Collective image—Montebello (middle-income Hispanic).

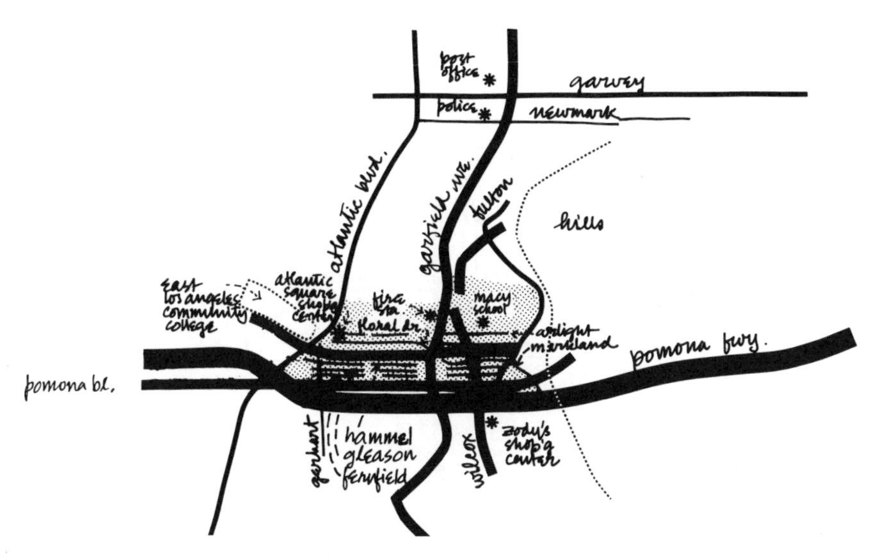

FIGURE 4.20. Collective image—Monterey Park (middle-income Hispanic).

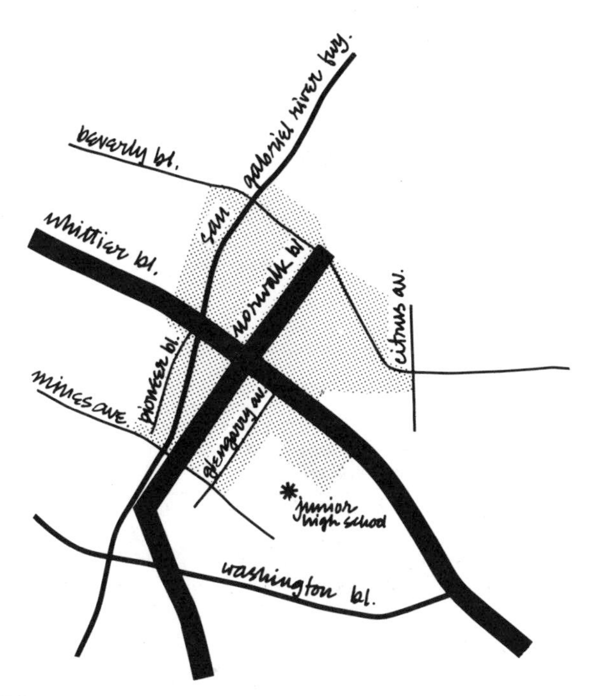

FIGURE 4.21. Collective image—Whittier (middle-income Hispanic).

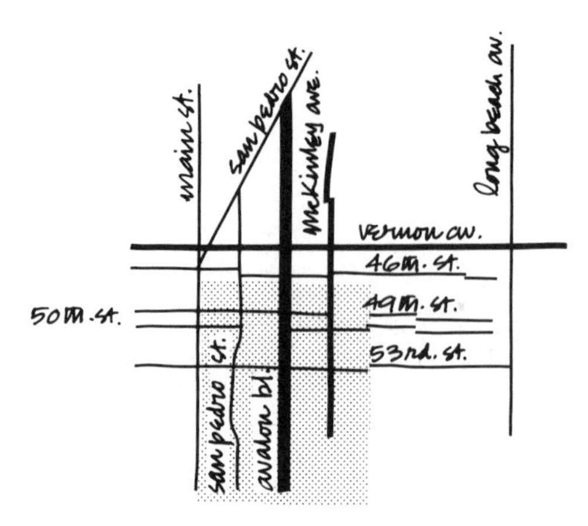

FIGURE 4.22. Collective image—Slauson (lower-income black).

puzzled as well. How is identity established? Because the question of place identity is important in the agenda of residential planning and design, we explored this question more directly by asking our respondents.[5] The responses to this question are summarized in Table 4.1. Although overall these findings seem to support what the collective images show, they also point to some important differences. For instance, schools—particularly elementary schools—showed up only as minor landmarks in the collective maps; yet, in response to our more specific inquiry about sources of place identity, they were the most frequently

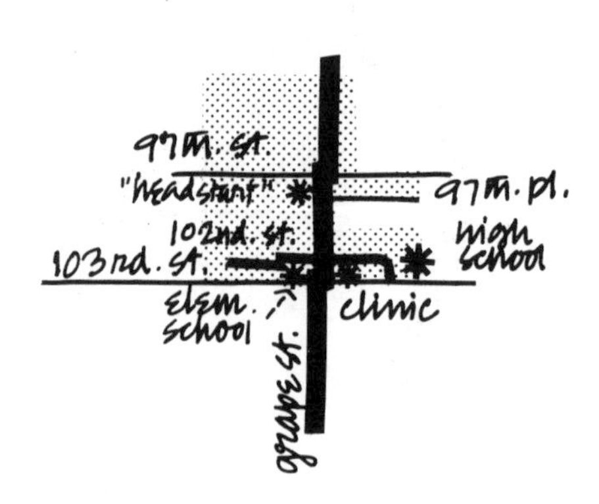

FIGURE 4.23. Collective image—Watts (lower-income black).

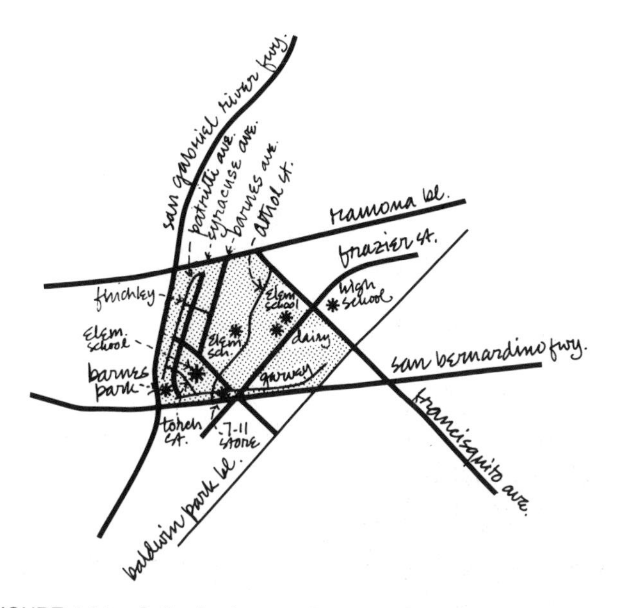

FIGURE 4.24. Collective image—Baldwin Park (lower-income white).

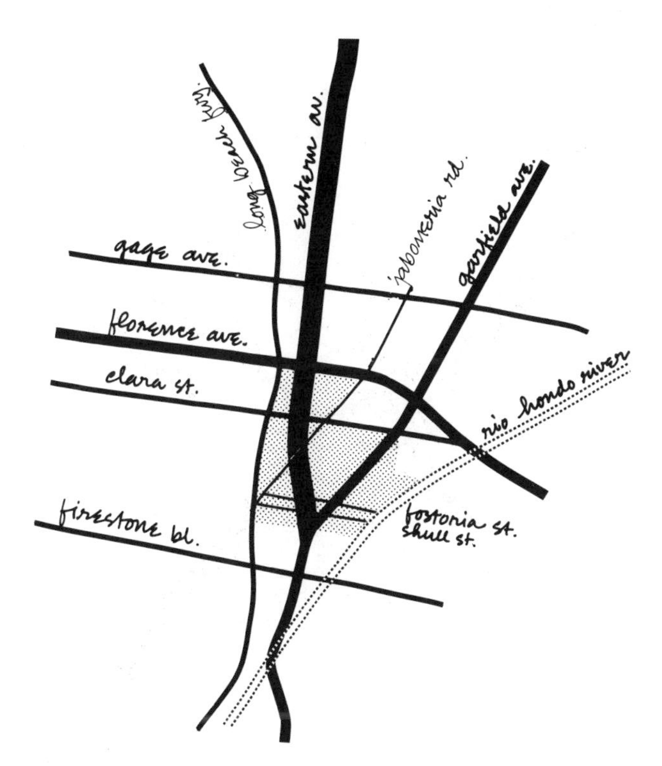

FIGURE 4.25. Collective image—Bell Gardens (lower-income white).

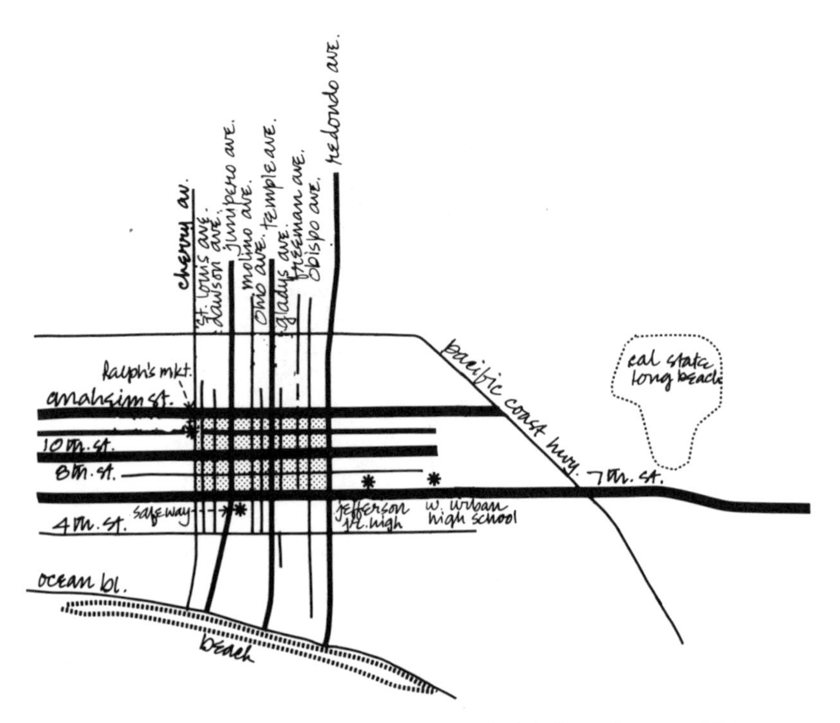

FIGURE 4.26. Collective image—Long Beach (lower-income white).

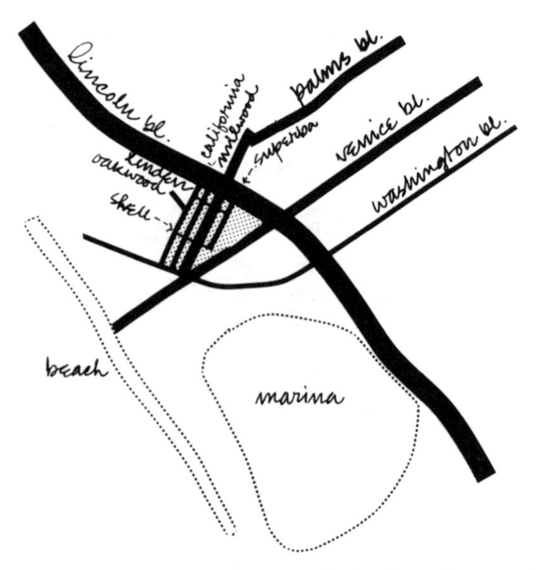

FIGURE 4.27. Collective image—Venice (lower-income white).

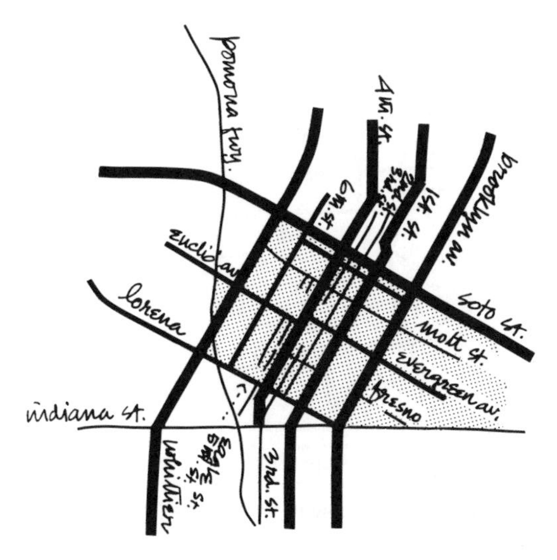

FIGURE 4.28. Collective image—Boyle Heights (lower-income Hispanic).

mentioned element. Streets, which were frequently depicted as organizing elements on the maps, showed up in the listing with increasing importance as income declined, although here they were associated with providing boundaries. It must be noted also that the sources of place identity were significantly different within the income groups, both in ranking and in number, indicating local differences. Moreover, the number of sources used to identify residential areas, on average, declined with income, neatly reflecting the decline in richness of pictorial representation with a decline in income as shown in the collective

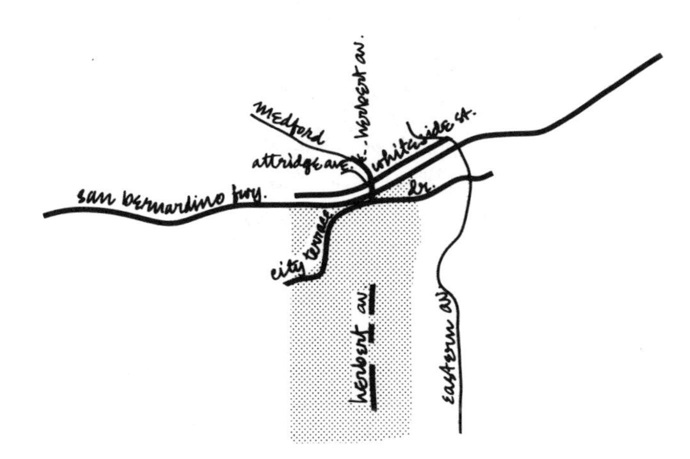

FIGURE 4.29. Collective image—City Terrace (lower-income Hispanic).

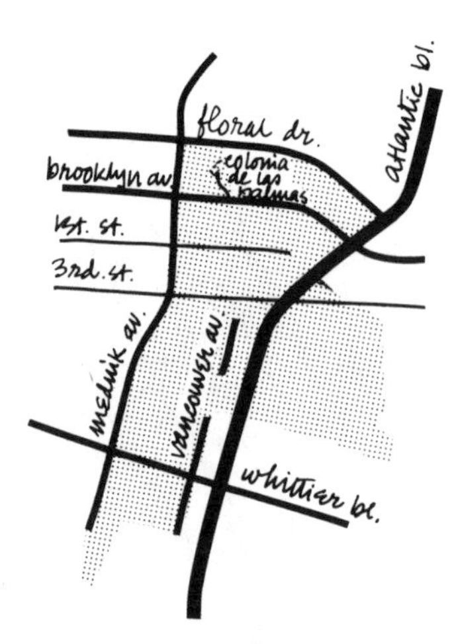

FIGURE 4.30. Collective image—East Los Angeles (lower-income Hispanic).

images (see the example, Figures 4.22, 4.25, 4.26, 4.27, 4.29, and 4.30). More on this in the following section.

DIFFERENCES IN COLLECTIVE IMAGES

The collective images of these residential areas varied from location to location in detail and content. In general, the maps drawn by the middle- and upper-income groups had a greater density of details than those of the lower-income respondents.[6] It is possible that the lack of details in the lower-income group maps was a reflection of the absence of distinctive and memorable features in these settings. It may also have been an indication of the antipathy toward and alienation from the milieu that were apparent in the residents' descriptions of the area (see Chapter 3).[7]

The upper-income-group maps, in contrast, were rich with embellishments. The opulence of environmental resources and amenities was clearly reflected in the frequent mentions of "tree-lined streets," "wooded areas," "bridle paths," "golf course," and the like. Mention of environmental resources was much less common in the middle-income-group maps. Instead, commercial establishments, local chain stores, and national franchise outlets were more commonly mentioned. The salient reference points in the middle- and some low-income-group maps were often local gas stations or drugstores. Indeed, chain stores, franchise names, and other corporate symbols were the main landmarks and embellishments in what otherwise appeared to be featureless landscapes.

TABLE 4.1. Sources of Residential Area Identity (Ranked by Population Groups)[a]

Upper-income white (n = 85)	Middle-income white (n = 80)	Middle-income Hispanic (n = 59)	Middle-income black (n = 86)	Lower-income white (n = 88)
		Residential area known by (ranked according to frequency)		
Schools (97.6)	Schools (97.5)	Schools (94.9)	Schools (96.5)	Schools (90.9)
Name (95.2)	Shopping area (82.5)	Shopping area (91.5)	Name (94.2)	Street boundary (81.8)
Natural or manmade features (88.1)	Housing type (77.5)	Street boundary (81.4)	Shopping area (89.5)	Name (62.5)
People (83.3)	Street boundary (75.0)	Name (66.1)	Street boundary (82.6)	Shopping area (62.5)
Shopping area (83.3)	Name (68.8)	Natural or manmade features (59.3)	Housing type (77.9)	Natural or manmade features (59.1)
Housing type (82.4)	Problems (68.8)	Housing type (57.6)	Natural or manmade features (68.6)	
Street boundary (82.1)	People (65.0)		People (64.0)	
Common problems (77.6)	Natural or manmade features (52.5)		Common problems (62.8)	
Others (50.6)				
Mean number of sources				
7.3	6.2	5.7	6.7	5.4

[a]Numbers within parentheses denote percentages. Only the items with frequencies of 51.0 percent or more are displayed. This questionnaire item was not included in the truncated version of the interview schedule for the lower-income blacks and Hispanics.

These maps varied from location to location in another aspect: the physical extent of the area shown. The contrast between the collective images of the Palos Verdes residents and those of the Watts residents was quite dramatic. The mean radius[8] of the Palos Verdes maps was over two miles; that of Watts was only one-eighth of a mile. Although these were extreme cases, the overall difference in the

TABLE 4.2. Average Size of Residential Area Maps

Residential area	n	Mean radius in miles[a]	Average for group
Upper-income white	85		
Pacific Palisades	17	0.62	
Bel-Air	23	1.09	
Palos Verdes	26	2.07	
San Marino	19	1.19	
			1.24
Middle-income white	80		
Westchester	20	1.88	
East Long Beach	20	0.86	
Van Nuys	21	1.06	
Temple City	19	0.85	
			1.16
Middle-income Hispanic	59		
Whittier	18	0.46	
Monterey Park	32	0.76	
Montebello	19	0.89	
			0.70
Middle-income black	86		
Carson	36	1.16	
Crenshaw	50	0.88	
			1.02
Lower-income white	80		
Venice	20	0.65	
Long Beach	21	0.64	
Bell Gardens	23	0.64	
Baldwin Park	24	0.39	
			0.58
Lower-income Hispanic	55		
Boyle Heights	26	0.52	
City Terrace	14	0.25	
East Los Angeles	15	0.40	
			0.39
Lower-income black	22		
Watts	13	0.13	
Slauson	9	0.19	
			0.16

[a]See note 8.

area covered in the maps of different groups followed a consistent pattern, as can be seen in Table 4.2. The residential area tended to shrink with declining income and to be smaller for minorities with the same income as whites. The distribution of the size of these maps is shown in Figure 4.31 where the ½ mile radius—the limiting dimension of the neighborhood unit—is used as the key benchmark. It is apparent that the residential area as conceptualized by a substantial majority of the upper- and middle-income respondents was larger than the maximum size prescribed in the neighborhood unit standards. For the majority of the low-income respondents, however, it was still well within this maximum dimension. Indeed, over a third of all low-income white and Hispanic respondents, and almost all of the black respondents, regardless of their income, showed their residential area as smaller than an area circumscribed by a quarter-mile radius.

This finding of the relationship between residential area size and income is so clear-cut that it warrants further comment. In another part of our interview, we asked our respondents about the activities in which they participated. We suspected that the activities that people engaged in were closely linked to their perceptions of their residential areas. Moreover, it is plausible to assume that the more activities one participates in, the more extensive one's residential area becomes. In all, our respondents were asked about their participation in eighty residentially related activities. The detail generated from this portion of the questionnaire permits only a summary here, so the activities have been aggregated into fourteen basic kinds, for example, personal care, shopping, and home maintenance. Table 4.3 shows each of the income and race/ethnicity groups and lists those basic activities in which each group either had the highest or the lowest participation rate. Although the table summarizes a considerable amount of

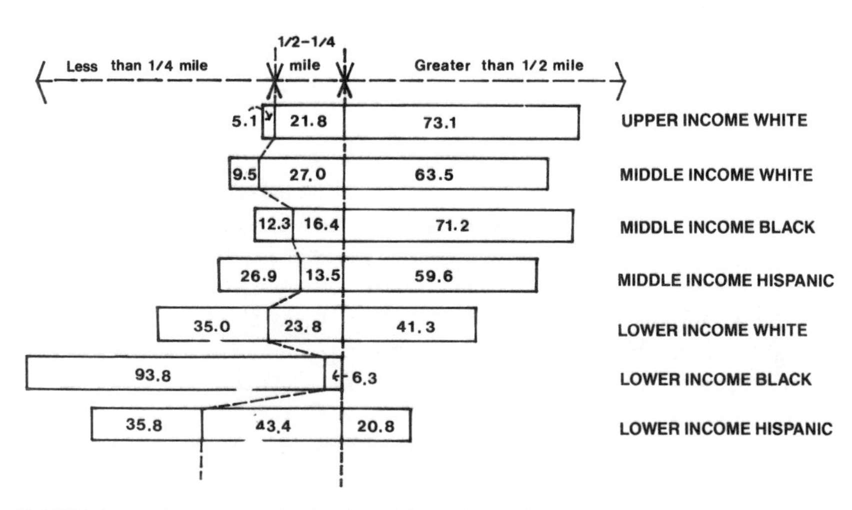

FIGURE 4.31. Percentage distribution of the radius of the residential area maps by population groups.

TABLE 4.3. Activity Participation Rate by Population Groups[a]

Population groups (*n*)	Highest participation (%)		Lowest participation (%)	
Upper-income white (85)	Out-of-home maintenance	87	Religious	47
	Cultural	84		
	Spectator entertainment	82		
	Shopping	77		
	Passive recreation	74		
	Social	64		
	Personal care	63		
	Active recreation	42		
Middle-income white (80)	In-home maintenance	82	Religious	47
	Passive reaction	74		
	Household miscellaneous	42		
Middle-income Hispanic (59)	Education	40		
	Participation sports	37		
Middle-income black (86)	Transportation-related	53		
Lower-income white (88)			Transportation-related	38
Lower-income Hispanic (55)			Out-of-home maintenance	33
			Personal care	46
			Shopping	65
			Social	52
			Household miscellaneous	18
			Participation sports	10
Lower-income black (22)	Religious	59	In-home maintenance	43
			Education	13
			Cultural	33
			Spectator entertainment	44
			Active recreation	10
			Passive recreation	55

[a]The percentage numbers denote the proportion of respondents in respective social groups participating in the activities of a given category.

information, it can be seen at a glance that a rough scale emerged: the upper- and middle-income white groups had the highest participation rates in ten of the fourteen activities, whereas the low-income black and Hispanic groups had the lowest participation in twelve of the fourteen activity groups.

We did not plot the location of each of the activities, so we cannot superimpose a set of the activities of the different groups on a composite map, but we believe that this evidence provides support for our hypothesis that activity and residential size are related, and that it is likely that the upper- and middle-income white groups perceive larger residential areas because they have a higher rate of participation, in general, in these residential-related activities.

This variability in the territorial range of the residential area and the activity participation rate of its inhabitants makes any attempt to prescribe an ideal size somewhat suspect. Furthermore, the apparent relationship between the size of the residential-area activity-participation rate and income or ethnicity raises the prospect that the residential area may be viewed differently by different social groups.

To be sure, the effects of physical form characteristics, density, land-use configurations, and so on cannot be ruled out in explaining the variability in the sizes of the residential area (Sims, 1973). Interlocality differences within a social group, in particular, can be attributed to local differences in site characteristics and development history.[9]

COMPARISON OF OUR FINDINGS WITH THE NEIGHBORHOOD UNIT CONCEPT

From the above, we see that not only are the constructs of environmental designers only partially realized in the minds of most people, but people also devise different organizing elements to conceptualize their residential area. For instance, whereas the neighborhood unit formula called for a series of largely self-contained cells surrounded by major arterials, our respondents conceived alternative constructs. Some saw the road system as a central spine, around which and to which items of interest were attached. Others saw the street system as a web or network, emphasizing the linking function of the streets connecting residential areas to other parts of the city, rather than the demarcating function of separating residential areas from one another.

Whereas the neighborhood unit emphasized major streets as visual and psychological boundaries or edges, we found our respondents more apt to note natural geographic and topological features as performing this function. And whereas the neighborhood unit construct had a strong emphasis on a central core, our respondents' images were usually devoid of this characteristic, tending at most only to hint at it and more often presenting several nodes of interest.

In this same vein, despite the designers' apparent preference that public facilities become the identifying landmarks of the neighborhood unit, we found that private establishments—particularly well-known chain stores—were equally salient to our respondents. Moreover, the regional shopping center, unknown in the days of the neighborhood unit's formulation, had become a strongly emphasized node. But its location is dictated by notions of residential purchasing power and share-of-the market considerations, which lie wholly outside the concepts of appropriate scale employed by Perry in his original neighborhood unit formulation.

Finally, the neighborhood scale was not the only organizing spatial unit used by our respondents. A district that encompassed many "neighborhoods," one or two regional shopping centers, and even a business area might be regarded as a residential area; sometimes political jurisdictions, such as adjacent cities and even a country, were included as well.

All in all, it would appear that, although the neighborhood unit is a useful organizing device in some instances, there is a far richer panoply of constructs that the designer can capitalize on and even improve on in the course of creating the residential environment.

Although the above themes were apparently independent of income, race/ethnicity, or stage in the family cycle, there were differences between these groups in other areas. Just as the verbal responses to our earlier queries were longer and usually more complete for the upper-income groups, their maps also contained more detail. Moreover, whereas the upper-income group's maps tended to stress environmental amenities and resources, the middle- and lower-income groups' maps were more likely to depict commercial establishments as distinguishing landmarks. This finding was supported in the listing of identifying features as well. Finally, the size of the area drawn was largest for the upper-income groups, and it decreased with income. And within a given income class, maps drawn by whites were larger than those drawn by nonwhites. All of these findings suggest possible differences in the alternative residential-area formulations that might be devised for each group, but they also raise some perplexing equity questions that we will address in the final chapter.

CONTRASTING THE RESIDENTIAL AREA WITH THE NEIGHBORHOOD

In all of the questions thus far, we asked the respondents to answer in terms of their residential area rather than their neighborhood, for we hoped not to contaminate their early responses by using the potentially loaded term *neighborhood.* On the other hand, at some point, we had to determine if there was any difference in their minds between the term *residential area* and the term *neighborhood.* We believed that the map-drawing portion of the interview was the easiest place in which to find out if these differences existed.

When asked if the neighborhood was of a different size than the residential area, roughly three-fifths of all respondents indicated that it was, and the rest regarded both areas as being the same size. Of those who suggested a two-tier conceptualization of the residential milieu, about one-third considered the neighborhood a larger entity, and the remaining two-thirds considered it smaller than the residential area. This distribution was generally similar for all population groups with the exception of the low-income blacks, who rarely (only one out of ten) made a distinction between the two concepts.

There are a number of possible explanations for these alternate conceptualizations of the residential environments. First, the responses could very well have been the result of an artificial distinction, an artifact of our asking the question in the first place. Second, the two-tier schema may have reflected a genuinely more complex cognitive organization than the coterminal schema. According to the former schema, the residential milieu may have been seen as consisting of an inner sanctum (whether termed a *residential area* or a *neighborhood)* and a larger, transitional—and perhaps interactive—zone. The fact that among whose who made a distinction between the two concepts about two-thirds showed the neighborhood as a smaller area suggests that the neighborhood, to most, symbolized an inner sanctum, whereas the residential area implied a larger orbit. Third, if the residential area was truly conceptualized as a sociospatial schema, as Lee (1968) has argued, the social and spatial components may not have been coterminous for some, although they were inseparable for others. That is, in the first case, the residential area may have meant a spatial concept and the neighborhood may have implied a social concept. In the latter case, this distinction may not have existed. Unfortunately, the data available to us do not shed further light on this matter. However, it is possible that the conceptualization of the residential milieu can be more complex than considered previously by the profession. This possibility poses some interesting challenges for the future organization of residential planning and design.

It will be recalled that the underlying values of the neighborhood unit concept discussed in Chapter 2 represented views of the intellectual and professional elite, as well as the views of the producers and lenders. It is not known how the public—the users of residential environments—perceived the necessity of neighborhood living fifty years ago. But in this study, we have been able to probe our respondents' views on the importance of neighborhood living by asking them the following questions:

> How important is it for you to live in a place that you consider a neighborhood? Is it (circle number)
>
> Very Important? 1 Fairly Important? 2 Not Important at all? 3
>
> Why do you say that?

In answering these questions, nearly nine out of ten respondents considered neighborhood living of at least some importance. Three out of five considered it very important. But as shown in Table 4.4, the white respondents from all income groups seemed to be slightly less intense about these feelings than their nonwhite counterparts. Whereas nine out of ten low-income blacks, seven out of ten middle-income blacks, and about two-thirds of the low- and middle-income Hispanics considered living in a neighborhood ''very important,'' only about half of the white respondents responded in a similar manner. Indeed, the percent-

TABLE 4.4. Importance of Living in a Neighborhood by Population Groups

Population groups	Very important (%)	Fairly important (%)	Not important at all (%)	n^a
Upper-income white	56.1	29.3	14.6	82
Middle-income white	50.6	39.0	10.4	77
Middle-income Hispanic	67.9	26.8	5.4	56
Middle-income black	73.8	20.2	6.0	84
Lower-income white	50.6	36.5	12.9	85
Lower-income black	90.9	9.1	0.0	22
Lower-income Hispanic	64.2	28.3	7.5	53

[a]Cases with missing data were not used to compute these percentages. Hence, the n's are somewhat smaller than those shown in most other tables.

age of respondents who did not consider living in a neighborhood important at all was much higher for all three white groups than for the nonwhite groups.

When broken down by stages in family cycle, elderly respondents and respondents form households with children showed similar preferences for neighborhood living (Table 4.5). Over 90% of these two groups felt that neighborhood living was of some importance, and as many as two-thirds of these respondents considered it very important. On the other hand, less than half of all respondents from households without children felt as strongly about neighborhood living. One out of five of these respondents did not consider neighborhood living of any importance at all.[10]

As we discussed previously, the neighborhood unit formula has been criticized as being a middle-income, white ideal and as being insensitive to the needs

TABLE 4.5. Importance of Living in a Neighborhood by Stages in Family Cycle

Stages in family cycle	Very important (%)	Fairly important (%)	Not important at all (%)	n^a
Households with children	64.5	29.0	6.5	248
Households without children	47.8	34.8	17.4	115
Elderly	69.8	22.9	7.3	96

[a]Cases with missing data were not used to compute these percentages. Hence, the n's are somewhat smaller than those shown in most other tables.

TABLE 4.6. Measures of Association (Kendall's Tau) between Importance of Living in a Neighborhood[a] and Personal Characteristics

Personal characteristics	First five[b] groups	All groups
Age	−.10**	−.11**
Education	.06	.08***
Family income	−.002	−.002
Race[c]	.18*	.20*
Housing tenure[d]	.09***	.09***
Sex[e]	−.10***	−.12**

[a]Importance of living in a neighborhood: 1 = very important; 2 = fairly important; 3 = not important at all.
[b]See Chapter 3, note 2.
[c]Race: 1 = nonwhite; 2 = white.
[d]Housing tenure: 1 = owners; 2 = renters.
[e]Sex: 1 = male; 2 = female.
*Significant at .001 level.
**Significant at .01 level.
***Significant at .05 level.

and ideals of other social groups; yet our data suggest that the "neighborhood" ideal not only is widely held among all population groups but may even be more coveted by nonwhite groups, who are more apt to feel left out of the "main-stream" white society. Families with children were expected to subscribe to the neighborhood ideal, but our data suggest that it is sought by the elderly as well.

Table 4.6 displays measures of association between the preference for neighborhood living and various personal characteristics. It is apparent that, among our respondents, age, race, housing tenure, and sex were all significantly associated with a preference for neighborhood living. This preference was likely to be held by older more than younger respondents, by nonwhites more than whites, by home owners more than renters, and by females more than males. Income did not appear to have any significant association with a preference for neighborhood living. Education was significant only when all seven groups were considered; the positive association suggests that those with a lower education level were more likely to suggest neighborhood living as important than were those with a higher level of education. These findings indicate a demographic profile of neighborhood dependency that is not inconsistent with the theoretical views and empirical findings reported by others (Suttles, 1973; Rainwater, 1966; Lee, 1968; Fried and Gleicher, 1961; Everitt and Cadwallader, 1972).

The stated reasons for the importance (or unimportance) of neighborhood living appeared to be both social and physical, as is apparent in Table 4.7 in which the most commonly cited reasons are listed along with examples of each. Here, we see a number of different facets of the social reasons that generally appeared to predominate. The need for human contact and a social network was the most important reason for wanting to live in a neighborhood. Despite different desires, virtually all our respondents saw the neighborhood as having a strong sociability aspect.

The next most common type of reason was the physical quality of the area. Here, there was a different type of disagreement: whereas those who liked neighborhood living saw it as less urban and more rural than other types of city living, those who did not want to live in a neighborhood found even the comparatively low density of neighborhood life still too crowded and "citified" for their satisfaction and comfort.

In a similar vein, family-related concerns were seen as important. For those who had children, the need for a sense of "neighborhood" was quite strong. For those without children, the sense of "neighborhood" was unimportant, but implicit in their comments was the recognition that it might have been important if they had a family and a different lifestyle.

Other concerns, such as the convenience of facilities, the personal and property safety of the inhabitants, the land use, and the rectitude of the other inhabitants, were mentioned as positive reasons for neighborhood living but were not mentioned by those who felt that neighborhood living was unimportant.

We would like to note here that these reasons can be viewed as value dimensions underlying the preference for neighborhood living, and that they very closely correspond to the various contextual, manifest, and tacit values discussed in Chapter 2. It will be recalled that the contextual values in the development of the neighborhood unit concept were essentially concerned with sociability, friendliness, the preservation of the family, the rectitude of the inhabitants, and livability. Interestingly, these values were echoed again fifty years later by the respondents in our study. The respondents also mentioned such physical values as convenience, shopping, personal safety, and privacy, all of which the promoters of the neighborhood unit concept advertised as its manifest values. Finally, we see in Table 4.7 mention of what can be categorized as tacit values (social homogeneity, property safety, and single-family homes) that were implicit in the endorsement of the neighborhood unit by various public and private institutions.

We next need to link these findings about perceptions of the "neighborhood" with our previous ones about the residential area. Ideally, we would have asked identical questions for each concept, but we wanted to avoid the tedium for the respondent involved in such an approach. In the main, we found that both concepts evoked responses with a social orientation rather than a physical one. Moreover, the differences in perception of the relative sizes of the neighborhood

TABLE 4.7. Most Commonly Cited Reasons for the Importance or Unimportance of Living in a Neighborhood

Importance of neighborhood living		
Rank according to frequency of mention	Reasons	Examples
1	Sociability	"Gives a chance to get involved with people. Helping each other with their needs."
2	Friendliness	"Who wants to be alone? Always lived in home. It's my nature. If you don't live in a neighborhood, you feel alone." "It's nice to be in a place where people are friendly. To me that's the most important thing about a neighborhood."
3	Quality of the area	"I consider it family living, gardens, flowers, trees, not quite as crowded as in the city."
4	Family-related	"For the children—environment." "Well, I think it is good for my children. There is a feeling of security. Their house is in a place that is easily identifiable. They have a place to call home." "If you have kids, it's nice for them to play with neighbor's kids."
5	Social homogeneity	"I think a neighborhood is a place where the same type of people are living—same income bracket." "I feel happier living among my own race."
6	Convenience	"I like the conveniences of being close to things like schools and shopping centers—I like to be able to walk where I want to go, especially since I don't have a car."
7	Shopping	"I'd like to be near food stores, drugstores, gas stations, garage." "I think of a neighborhood as a residential area with a small area for stores such as cleaners, drug, grocery store."
8	Personal safety	"Safe for children to play so that they won't be picked up or run over or molested. Safe for old ladies."
8	Property safety	"I live alone with my grandson, and I feel more secure living in a neighborhood rather than in an isolated area. We have police patrol in our area, which I like."
9	Single-family homes	"Don't want apartment or commercial—just want residential (single-family) area."
9	Moral	"Because you want to live in a neighborhood

(*continued*)

TABLE 4.7. (*Continued*)

Rank according to frequency of mention	Reasons	Examples
		where you can bring up your kids, have a decent area, be proud of your neighborhood and residents. Give to your neighborhood to keep it growing."
10	Quiet	"Because in a neighborhood, there is not too much noise."
		"Because a residential neighborhood is far from industry. There is not the same amount of noise as in other districts."
	Unimportance of neighborhood living	
Frequency not ranked	**Reasons**	**Examples**
a	Sociability/friendliness	"We really could care less who our neighbors are; never socialize with neighbors; not home, anyway."
b	Urban/rural	"Because I'm a nature fiend and could live all by myself."
		"Because I'd be very happy to live in a rural area."
		"Unimportant, very unimportant. Because I like open space. I like pretty open fields."
c	Family-related	"Because I don't have any kids."
		"Don't have a family, not here that much of the time. I'm mostly out or away somewhere else."
		"Not very important—we no longer have young children; our interests are far, far away from home—many times in other parts of the world."
d	Privacy	"Enjoy privacy. Growing up, most of my family in a tract home, I was made aware of some of the disadvantages of living in a neighborhood."
		"I would really like to live in a place where there aren't many people, like out in the mountains."
		"I want to live in the woods; I'd like to be a hermit."

and the residential area suggested that there could not be a great difference in the attributes of the two, even if they were perceived as distinct.

Accordingly, we are of the view that our findings for residential areas apply pretty much to separate notions of the neighborhood as well. Certainly, the variety of basic concepts that we have uncovered would be applicable to design paradigms for different residential environments. Moreover, there appears to be nothing sacred about the neighborhood unit formula *per se* as an ideal solution—even to our respondents who conceived of a neighborhood as being distinct from a residential area.

Finally, it would appear that the notion of the neighborhood, whatever physical form it might take, is strongly fixed in many people's minds as being in some way important to their residential well-being. Whatever formulas are devised that differ from the neighborhood unit concept, they should offer alternatives to it rather than its replacement, for whatever the shortcomings of the neighborhood unit, it may serve some purpose in meeting people's needs in a number of areas.

SUMMARY AND CONCLUSIONS

In this chapter, we have explored residential settings by examining individual and collective images of the residential area. As in the previous chapter, we have searched for overall patterns in people's perceptions of that residential area. In the process, we have also explored which elements of the physical environment are critical in shaping the residential image as well as the extent to which these elements have matched the planner's traditional preconceptions. Finally, we have tried to establish the relationship between the residential area and the neighborhood, and to examine the perceived importance of neighborhood living.

Some general patterns have become apparent. It is possible to distinguish four or five general types of images from the maps constructed by our respondents. These types suggest that the residential area is seen as an activity node, as a "face-block" intimate community, as a behavior circuit, as a "defended neighborhood," and as a "nonplace" of limited liability. But these are only tentative inferences, for many maps are a combination, and still others defy any categorization.

The collective images reveal some additional facets of the physical-spatial aspects of the residential area. The street system is the major organizing element. Arterials or highways were used consistently by our respondents as the major frame of reference; their maps were either centrally organized around a main "spine" or were noncentrally organized as a "net" or a "grid." The street played a particularly significant role where the residental setting lacked any distinctive natural or manmade elements.

Moreover, the images of the residential areas reflect qualitative differences in the settings, consistent with the pattern seen previously in the verbal descriptions of these areas. In general, the maps of the upper-income areas depicted the opulence of their environmental resources and amenities and the generous supply of public services and facilities in those areas. In some middle-income areas and in all low-income areas where such resources did not exist, large commercial establishments or the corporate symbols of chain stores became the important distinguishing features of the residential landscape. Typically, these institutions are the villains who spoil the planner's dream environment. However, it is apparent that in the visual environment of everyday life, people not only accept these commercial landmarks but depend on them for organizing their city experiences.

Admittedly, none of these areas was deliberately designed according to neighborhood unit principles. But as suggested by Solow *et al.* (1969), many of these principles and criteria have been absorbed by the zoning ordinances and subdivision regulations. Indeed, the objective environments of most of our interview areas, although not reflecting the textbook model, included many of the essential features of the neighborhood unit concept: schools, parks, community shopping centers, neighborhood services and facilities, and arterials bounding the area. But very few, if any, of the collective images of our respondents resembled the ideal of the neighborhood unit. The sense of boundary was usually absent, and so was a sense of a center. Rather than a center and a perimeter, a linear or a grid reference system was the hallmark of these images.

The possibility that the residential environment implies a two-tier schema, having noncoterminous social and spatial components, and being organized according to a hierarchy of personal significance, puts the familiar neighborhood unit in a dubious light.

Nevertheless, the larger notion of neighborhood living was considered important by a substantial majority of our respondents, especially the nonwhites. In general, older more than younger respondents, owners more than renters, and females more than males were likely to consider neighborhood living important. But whether our respondents preferred to live in a neighborhood—or preferred not to—the reasons given were largely the same. Social, rather than physical, reasons were most commonly offered.

A NOTE ON METHODOLOGY AND FINDINGS

It is appropriate here to comment briefly on the methods used in Chapters 3 and 4, and on the findings they produced. It seems clear that what one asks, as well as the medium in which the answer is sought, shapes the substance of the response. For instance, when we asked people to describe the residential area verbally, they responded with a social rather than a physical focus. Later, when we asked them to draw their residential area, they responded with a physical rather than a social focus.

Which is the reality? We note (without comment) that sociologists use words as the primary medium of expression in their profession, whereas environmental designers use diagrams. Sociologists have found social relations paramount in residential existence; designers have traditionally professed a belief in the importance of the physical environment. It appears that our respondents provided a pale reflection of the beliefs of each profession when they were asked to use that profession's medium of expression. Perhaps just as there is a sociology of knowledge dependent on the background and the training of the investigator, there is also a technology of knowledge, dependent on the medium that the investigator chooses to elicit responses.

Which is more important—the social or the physical? We have chosen to side with the social, but then, our survey was more dependent on the use of words than on drawings, for surveys are word-oriented. Interestingly, when a third approach was used—that is, when we asked our respondents to rank a list of preselected items for their importance in providing the identity of a residential environment—the responses were again different. Here, we found that the item "people," which so strongly showed up in our open-ended question, were ranked in the middle. "Streets," which were emphasized in the collective cognitive maps were also ranked in the middle of this list. It was the item "schools," which showed weakly in the other two queries, that was the most important in the ranking.

We can conclude only that research on this topic truly requires a variety of approaches, no one of which will provide a definitive answer to these kinds of questions. On the other hand, apropos of our own research, we wish to point out that, regardless of the forms of inquiry and response, income consistently showed as an important variable.

NOTES

1. For a definition of the term *action space,* see Wolpert (1965) and Horton and Reynolds (1971).
2. Webber (1964) argued that the importance of place and the local community diminishes with increasing income and education. The upwardly mobile professional class are likely to belong to many different "realms" based on their professional and class ties, the least likely being a "place" - bound realm. Webber conceded however, that among the lower-income groups, the local community and the place may still play an important role in friendship formation and as a focus of social interaction.
3. The concept of *home area* has been discussed by Everitt and Cadwallader (1972) and is defined as "an area of importance of significance around the home—that might be comparable from

person to person. Thus the home area of an individual is the area around his house in which he feels most at home'' (p. 1-2-2).

4. The order to prepare the ''collective image'' of an area, we kept a simple tally of the various items (both labeled and unlabelled) shown in the maps. The items shown in the individual maps were coded into the following categories: streets, highways, and freeways; public facilities (e.g., police stations, fire stations, and post offices); private facilities (e.g., drugstores, supermarkets, and gas stations); natural amenities (e.g., beaches, mountains, and rivers); and districts (e.g., place names, campuses, ''residential,'' and ''business district''). The ''composite'' residential area maps for each locality were prepared by transferring the items mentioned in the maps drawn by the respondents to an overlay of a standard street map, and then by graphically differentiating the frequencies with which they were mentioned.
5. See Question 11, Appendix I.
6. With the exception of Long Beach, all low-income area maps had, on the average, less than eleven items per map, whereas the others had thirteen or more.
7. It may also reflect a fundamental lack of enthusiasm on the part of the low-income respondents in drawing these maps. Because this possibility cannot be ruled out, any inference about the role of the objective environmental conditions or the residents' attitude toward the place can be made only with caution.
8. The radius of a map was computed by redrawing the map (as well as it could be interpreted) on a tracing-paper overlay over a standard street map drawn to scale. Once the streets shown in the original were outlined in the overlay, a circle was drawn to circumscribe the entire map. The radius of this circumscribing circle is the measure referred to as the *radius* of the residential area maps. The unit of measure was one mile.
9. Consider, for example, the difference between the maps drawn by residents of Palos Verdes and those drawn by residents of Pacific Palisades, both upper-income white areas. The average radius of the Pacific Palisades map was one-third that of the Palos Verdes maps. In this case, the difference may very well have arisen from physical differences in the settings. The Pacific Palisades setting is well delineated by a system of highways, streets, canyons, and palisades. Most respondents drew consistent boundaries that delineated an area much smaller than the area that the Palos Verdes respondents drew. In their case, the actual interview area lacked distinctive boundaries, but the larger context of the Palos Verdes peninsula is extremely well-defined because of its encircling coastline, hilly terrain, low-density development, and expensive homes. Additionally, the interview area shared the uniformity and exclusivity of appearance with most of the peninsula that grew out of the strict land-use, density, and architectural controls applied throughout the peninsula. Thus, it is no surprise that the Palos Verdeans projected an expansive concept of the residential area that included the entire peninsula.

   It is also possible to make a theoretical argument that the size of the residential area may reflect the objective density of the area. To the extent that cognitive maps are shaped by the locus of various residential activities (e.g., shopping for food, taking a walk, and visiting a neighbor), and to the extent that density dictates the spatial concentration of public or private services and facilities, some association can be expected between the residential area density and the size of the cognitive maps.
10. To test these relations further, we cross-tabulated the importance of neighborhood living by collapsing the population groups into white and nonwhite categories, and by collapsing the elderly and the families with children into one category. The partial gamma for the association between neighborhood living and population groups (1 = white; 2 = non-white) was −.36 (and −.37 when only the first five groups were taken), when the stage in the family cycle was controlled for. The partial gamma for the association between neighborhood living and the stage in the family cycle (1 = elderly or families with children; 2 = families without children) was .37 (and also .37 when only the first five groups were considered), when the population groups were controlled for.

5

The Residential Area as a Physical Place

The Setting

The impressions, evaluations, and images of the residential area discussed previously clearly suggest the importance of the physical place, as well as the preeminence of the social milieu. The residential area as a physical place is of particular interest to planners and designers because physical layout, housing mix, the composition of land use, and environmental furnishings are still largely determined by their professional judgment. These are their areas of expertise. It is apparent from our respondents' mental maps as well as from their collective images that the residential area is most commonly represented as an area of housing (including the individual home) surrounded by or mixed with several "nonresidential" uses of the environment. For future residential planning and design, it would be advantageous to know which of these environmental elements or land uses are acceptable, if not desirable, in the residential context, as well as ones which are clearly unacceptable. If it can be shown that the presence or the absence of particular elements or land uses is linked to residential satisfaction and well-being, significant progress can be made toward establishing some criteria for the design of a "good" residential environment for all.

In general, the psychological and behavioral implications of using land in different ways have never been explicitly considered in the making of land-use decisions. Indeed, the common descriptions of land use—industrial, commercial, residential, and institutional—reflect their economic and social functions rather than their behavioral characteristics, which the term *use* implies (Wohlwill, 1975). Health, safety, and nuisance considerations have typically been invoked in resolving land-use controversies, but the perceptual and behavioral impact of the resultant land use policies has rarely been considered. However, the lessons of urban renewal and freeway development have taught us that perceptual and behavioral impact is equally important in land-use decisions. There is a growing opinion that the mix and the spatial structuring of land use affects not only people's health and safety but also their perceived quality of life and their sense of well-being.

We suggest that the perception of specific land uses is likely to be shaped by their relevance, utility, and meaning to an individual. To some, a Laundromat may be a basic necessity in the residential area; to others, it may be seen as drawing transient or lower-class customers and socially undesirable people. For a gregarious imbiber, a tavern or a liquor store is a convenience; for a teetotaler, both may symbolize decadence. Alternatively, the perception of a particular land use may reflect its impact on the senses or on safety and security. Thus, for some, an amusement park in the residential area is too noisy, or a baseball park generates too much traffic.

But if behavioral perspectives are to be included in land-use planning-principles, conventional typologies of land use may be inadequate. It may be necessary to develop constructs of environmental elements in a more individual-behavioral sense (Wohlwill, 1975). Thus, neighborhood parks, drugstores, gas stations, department stores, movie theaters, and the like may be considered more relevant descriptors of land use than such conventional categories as residential, commercial, industrial, and so on. From an individual-behavioral standpoint, a "drugstore" is more than just a type of "commercial" land use, it is a "behavior setting" as well (Barker, 1968). Indeed, limited studies of environmental cognition suggest that the way we catalog and organize our everday experience in memory and thought, and in language and conversation, does not follow the scheme of conventional land-use categories (Lowenthal, 1972; Carr and Schissler, 1969). Rather, these categories are based on such common descriptions as "supermarkets," "auto repair shops," "tot lots," and the like, which are common in everyday language.

In our attempt to define the physical composition of the residential environment, we have chosen to use the term *environmental settings* to define most of the features that can be part of one's residential area. Although this term is clearly inspired by the theoretical construct of the *behavior setting,* as defined by Barker (1968), we do not intend to use this concept for its exact original purposes (as in studies of local behavior), or with the same level of precision.[1] For the purposes of our study, it will suffice to look at, for example, a stationery store or an arcade simply as an "environmental setting," without concerning ourselves with what the "standing rules of behavior" or the "synomorphs" of the setting and the behavior are. By the term *environmental setting,* we mean spaces,

FIGURE 5.1. A setting for sidewalk activity in Boyle Heights (lower-income Hispanic).

facilities, establishments, and so on, either privately or publicly owned, that are used for one or more activities in a reasonably predictable and recognizable manner. We further recognize the possibility that certain environmental elements that do not strictly qualify as ''settings'' for activities can, nevertheless, contribute indirectly to the satisfactory performance of an activity or to one's overall feeling of satisfaction and well-being. Signs, billboards, and utility lines are examples of environmental objects that are not settings but that may significantly affect one's satisfaction with the environment. These objects are called, for lack of a better term, *environmental hardware*.[2] (See Figures 5.1–5.9.)

Thus, we propose that the residential area as a physical place can be seen as consisting of one's home and other dwellings, plus a constellation of environmental settings and hardware. To the extent that the existing situation matches one's concept of desirable settings and hardware, it will be perceived as satisfactory. Incongruence will result in stress and dissatisfaction. But more significantly, from looking at the residents' preferences for settings and hardware and at the match or mismatch in specific environments, it may be possible to define not only criteria for good residential environments but also priorities and strategies for improving the conditions of different social groups. What follows is a

FIGURE 5.2. The front yard as a children's activity setting—a scene from City Terrace (lower-income Hispanic).

discussion of people's perceptions of and preferences for various settings and hardware based on one segment of our data.

DESIRABLE SETTINGS AND HARDWARE

In the course of the interview, our respondents were given a deck of 78 cards with a description of a setting or an item of hardware written on each (Table 5.3 gives the complete list).[3] The respondents used two sets of these cards to indicate, first, their perception of their current residential environment and, second, how they would like it to be. They sorted the first set into three piles that were labeled (1) "things actually in my area"; (2) "things actually out of my area"; and (3) "don't know." They then sorted the second set into three piles, which were this time labeled (1) "things I *want* in my residential area"; (2) "things I *want out* of my residential area"; and (3) "either in or out is fine." Thus, the first card-sort was designed as a measure of neighborhood cognition, and the second was a measure of preference for the ideal setting. Collectively, these two sets of data provided a basis for assessing the level of congruence between the existing and the desired. Let us first examine which environmental settings and hardware were typically desired by people in their residential area and what, if any, differences existed between different population groups and stages in the family-cycle.

FIGURE 5.3. A pedestrian overpass in Boyle Heights (lower-income Hispanic)—an environmental setting or environmental hardware?

FIGURE 5.4. Local stores were commonly desired settings—Venice (lower-income white).

FIGURE 5.5. A school play-yard (an acceptable setting)—in Westchester (middle-income white).

FIGURE 5.6. Sidewalk settings in Long Beach (lower-income white).

In Figure 5.10 we present a list of the environmental settings and hardware desired ("I *want* in my residential area . . .") by the majority (at least four out of seven population groups).[4] Interestingly, the list was quite long and included 58 of the 78 items in the deck of cards. A closer look at the table reveals that the desirability of most of the items was not uniform for different population groups and that within a given population group not all elements were desired with equally strong intensity. The items are ranked according to their overall desirability (based on the frequency of mention) for all seven population groups. The first fourteen of these elements represent a core of settings and hardware that were strongly desired by the majority of the groups. The next twenty-two elements represent preferences of moderate intensity, and the remaining twenty represent mostly weak preferences—and for some groups, no majority preferences at all.

FIGURE 5.7. Outdoor benches and bus stops were seen as acceptable settings—an example from Boyle Heights (lower-income Hispanic).

FIGURE 5.8. A billboard in the front yard! Undersirable environmental hardware in Venice (lower-income white).

FIGURE 5.9. Utility poles in Bell Gardens (lower-income white)—another example of undesirable environmental hardware.

The first fourteen items on this list are seemingly the least controversial settings and hardware to be included in a residential area. A library, a neighborhood park, a fire station, bus stops, a post office, street lights, walkways and pedestrian crossings, and an elementary school appear to be the core public facilities and amenties in a desirable residential place, echoing the theme of the neighborhood unit concept. A careful perusal of *Planning the Neighborhood* will reveal that food- and drugstores are indeed included in the list of their neighborhood shopping facilities, as are medical and dental services (APHA, 1960). Gas stations and banks or savings and loans are not mentioned specifically but could easily have been subsumed under the category of "miscellaneous services" prescribed by this earlier book of standards. Thus, it seems that in the case of neighborhood values as discussed in Chapter 4, public perception of what constituted the core environmental settings and hardware of the residential area

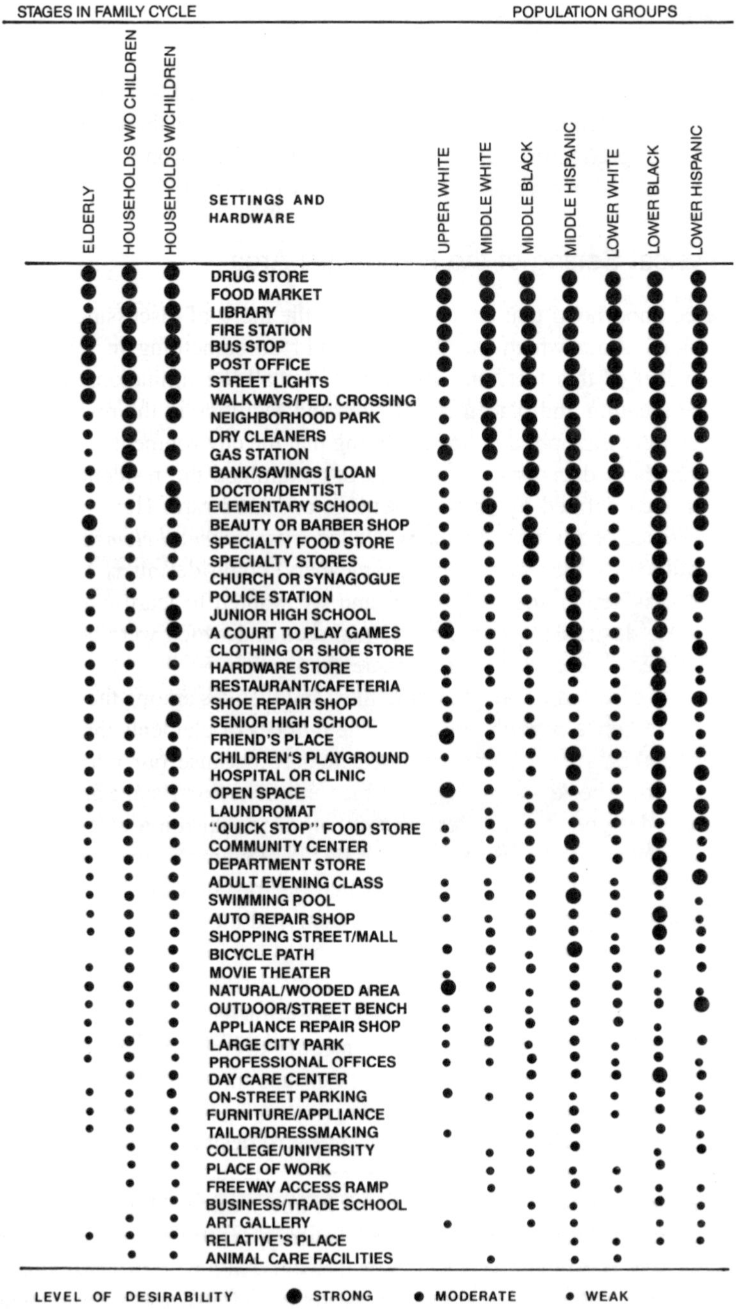

STAGES IN FAMILY CYCLE | POPULATION GROUPS

ELDERLY | HOUSEHOLDS W/O CHILDREN | HOUSEHOLDS W/CHILDREN | SETTINGS AND HARDWARE | UPPER WHITE | MIDDLE WHITE | MIDDLE BLACK | MIDDLE HISPANIC | LOWER WHITE | LOWER BLACK | LOWER HISPANIC

DRUG STORE
FOOD MARKET
LIBRARY
FIRE STATION
BUS STOP
POST OFFICE
STREET LIGHTS
WALKWAYS/PED. CROSSING
NEIGHBORHOOD PARK
DRY CLEANERS
GAS STATION
BANK/SAVINGS [LOAN
DOCTOR/DENTIST
ELEMENTARY SCHOOL
BEAUTY OR BARBER SHOP
SPECIALTY FOOD STORE
SPECIALTY STORES
CHURCH OR SYNAGOGUE
POLICE STATION
JUNIOR HIGH SCHOOL
A COURT TO PLAY GAMES
CLOTHING OR SHOE STORE
HARDWARE STORE
RESTAURANT/CAFETERIA
SHOE REPAIR SHOP
SENIOR HIGH SCHOOL
FRIEND'S PLACE
CHILDREN'S PLAYGROUND
HOSPITAL OR CLINIC
OPEN SPACE
LAUNDROMAT
"QUICK STOP" FOOD STORE
COMMUNITY CENTER
DEPARTMENT STORE
ADULT EVENING CLASS
SWIMMING POOL
AUTO REPAIR SHOP
SHOPPING STREET/MALL
BICYCLE PATH
MOVIE THEATER
NATURAL/WOODED AREA
OUTDOOR/STREET BENCH
APPLIANCE REPAIR SHOP
LARGE CITY PARK
PROFESSIONAL OFFICES
DAY CARE CENTER
ON-STREET PARKING
FURNITURE/APPLIANCE
TAILOR/DRESSMAKING
COLLEGE/UNIVERSITY
PLACE OF WORK
FREEWAY ACCESS RAMP
BUSINESS/TRADE SCHOOL
ART GALLERY
RELATIVE'S PLACE
ANIMAL CARE FACILITIES

LEVEL OF DESIRABILITY ● STRONG • MODERATE · WEAK

FIGURE 5.10. Levels of desirability for different environmental settings and hardware by population groups and stages in the family cycle. (Strong desirability = 83.4% or more wanting; moderate desirability = 66.7–83.3% wanting; weak desirability = 50.1–66.7% wanting.)

have remained unchanged despite shifts in lifestyle and activity patterns. However, once we reach the level of the elements of moderate desirability range, the intergroup differences become more and more apparent. For example, such items as hospital or clinic, Laundromat, and department store were not in the "wanted" list of the upper-income whites, whereas some of the lower-income groups displayed strong interests in them. The consensus among groups declines rapidly as we reach the bottom tiers of environmental settings and hardware.

Some overall patterns are obvious. It is clear that the upper-income whites were the most exclusive in their concept of residential area, and that the two lower-income minority groups were the most inclusive. This is an anomalous result in light of the large residential areas depicted by the upper-income whites and the small areas shown by the lower-income minority groups (see Chapter 4). But no clear pattern by income is obvious because the middle-income Hispanics also appeared to be as inclusive as the lower-income Hispanics or blacks. The low-income whites and the middle-income blacks, on the other hand, appeared to be slightly less inclusive, and the middle-income whites appeared to be quite exclusive—next only to the upper-income whites. The evidence is still too slim to make any confident generalization about ethnic differences in preferences for these elements of secondary and tertiary importance, but it appears that the whites generally were more exclusive in their land-use perferences than their minority counterparts.[5]

Differences in the desired settings and hardware by stage in the family cycle are also shown in Figure 5.10. Once again, for top-ranked items, a great deal of consensus existed between stages in the family cycle but declines for the bottom tier of elements. The elderly appeared to be the most exclusive of the stages in the family cycle. In fact, for many of the items, the elderly showed only a weak preference, whereas the other two stages in the family cycle showed moderate or even strong preferences. It seems obvious that physical incapacities, retirement living, and limited physical mobility made many of the environmental settings matter less to the elderly. Thus, for example, a place of work, a day-care center, or a gymnasium is no longer a part of what elderly persons would define as their relevant action space.

UNDESIRABLE SETTINGS AND HARDWARE

Relatively fewer elements were *not* wanted in the residential area. This is apparent from the list of "unwanted" elements for different population groups shown in Table 5.1. Although our discussion here is limited to the elements not desired ("wanted out") by 50 percent or more of the respondents from each group, we have also included in this table elements not wanted by at least a third of the respondents. (We believe that these elements are also of potential interest to the planner, even though they are not necessarily "wanted out" by a clear majority.)

TABLE 5.1. List of Environmental Elements "Wanted Out" of the Residential Area

Upper-income white ($n = 85$)	Middle-income white ($n = 80$)	Middle-income Hispanic ($n = 59$)	Middle-income black ($n = 86$)	Lower-income white ($n = 88$)	Lower-income Hispanic ($n = 55$)	Lower-income black ($n = 22$)
Amusement park	Utility lines	Billboards	Utility lines	Billboards	Billboards	Liquor store
Utility lines	Billboards	Alleys	Billboards	Utility lines[a]	Alleys	Night club
Billboards	Alleys	Marina[a]	Alleys	Bar[a]	Bar	Alleys
Night club or discotheque	Amusement park[a]	Utility lines[a]	Amusement park[a]	Night club[a]	Liquor store[a]	Zoo
Bus terminal	Bar[a]	Zoo[a]	Marina[a]		Thrift shop[a]	Beach
Bowling alley	Night club[a]		Bar[a]			Club or lodge
Zoo	Bus terminal[a]		Night club[a]			Marina[a]
Sports arena	Thrift shop[a]		Bus terminal[a]			Utility lines[a]
Business or trade school	Zoo[a]		Zoo[a]			Bus terminal[a]
Freeway access ramp	Sports arena[a]		Sports arena[a]			
Marina[a]			Beach[a]			
Bar[a]						
Thrift shop[a]						
Skating rink[a]						
Alleys[a]						
Department store[a]						

[a]Elements that were "wanted out" by a third or more, but less than a half, of the sample groups. The remaining elements were "wanted out" by at least half of the respondents.

Once again, it is apparent that the upper-income whites were much more exclusive in their tastes for environmental elements than were the rest of the respondents. Many of the elements not wanted by this group were clearly facilities that were regional in scope and were potential sources of traffic, congestion, and other social nuisance. Amusement parks, bus terminals, zoos, sports arenas, bowling alleys, and so on were rejected by the upper-income groups.

Their dislike of some of the other settings and hardware was shared by other population groups as well. For example, alleys, billboards, and utility lines—some of the most ubiquitous elements of the inner-city (and older suburban) environment—appeared to be the most common nuisances, although all low-income groups and middle-income Hispanics seemed to be more tolerant of overhead utility lines. Low-income blacks did not seem to mind even the billboards, but they did not want bars, nightclubs, liquor stores, or a club or a lodge in their area.[6] In enumeration by stages in the family cycle (Table 5.2), both billboards and utility lines stand out as the consensus choice of undesirable elements. Both belong to the category of environmental hardware and are related to the perceptual quality of the environment. Households with children also showed some aversion to alleys—stemming possibly from a concern about the safety and the security of children. The elderly did not want a discotheque, a nightclub, a bar, or a cocktail lounge in their area. Most probably their aversion to these settings resulted from perceptions of the crowd that these settings may draw and the resultant noise, traffic, and reckless behavior. Other items in this

TABLE 5.2. List of Environmental Elements "Wanted Out" of the Residential Area by Stages in Family Cycle

Households with children (n = 256)	Households without children (n = 117)	Elderly (n = 102)
Billboards	Billboards	Billboards
Utility lines	Utility lines	Utility lines
Alleys	Alleys[a]	Discotheque or night club
Amusement parks or fairgrounds[a]	Sports arena or stadium[a]	Bar or cocktail lounge
Discotheque or night club[a]	Zoo[a]	Zoo[a]
Zoo[a]	Discotheque or night club[a]	Alleys[a]
Bus terminal[a]	Amusement parks or fairgrounds[a]	Marina or boat dock[a]
Bar or cocktail lounge[a]	Bar or cocktail lounge[a]	Sports arena or stadium
Marina or boat dock[a]	Bus terminal[a]	Liquor store[a]
Sports arena or stadium[a]	Marina or boat dock[a]	Beach[a]
	Thrift shop or secondhand store[a]	

[a]Elements that were "wanted out" only by a third, but less than a half, of the sample groups. The remaining elements were "wanted out" by at least half of the respondents.

table are quite similar to the ones displayed in Table 5.1, and although they were not "unwanted" by a clear majority, a substantial minority both in terms of population group and stages in the family cycle considered them undesirable (see Figures 5.4, 5.7, 5.8, and 5.9).

So far, we have discussed whether these environmental settings and hardware were or were not wanted by the majority of our respondents in a particular group. We have implied, in our ranking of these items by the frequency of their mention, a certain ordering of salience—a sense of the intensity with which a particular setting or a piece of environmental hardware was desired or not desired. Yet, this display tells only a partial story; it does not include information about the extent to which the remaining respondents were negative or neutral. Nor for that matter does it include an ordering of these environmental settings and hardware based on the aggregate responses from all population groups and all stages in the family cycle. Tables 5.1 and 5.2 were designed to be of use to planners and designers involved in designing the residential area for a particular social group, and the enumeration of the elements by social groups is expected to provide a useful, particularized checklist. We also felt that, as a matter of public policy affecting *all* segments of the population, a composite index might be more useful operationally for a planner who is about to rewrite zoning ordinances or subdivision regulations for a city. Accordingly, we have developed composite scores for these environmental elements that place them in a continuum of values ranging from +1 to −1, based on the average of the total number of positive ("I want in my area"), indifferent ("either in or out is fine") and negative ("I want out of my area") votes received by each element.[7] Table 5.3 shows the location of various environmental settings and hardware elements on a residential desirability scale ranging from extremely desirable to extremely undesirable. This scaling of the elements can be seen as a metric of the psychological distance of a particular setting from one's home and from other settings, as well. Additionally, this scale identifies the "neutral" elements, about which people are generally indifferent or ambivalent; the overall presence or absence of these elements may not contribute much to overall residential well-being. Finally, this desirability scale can be used to evaluate operationally the physical place qualities of an area and to estimate the "intensity" of setting deprivation or aggravation, which are discussed in the following section.

SETTING DEPRIVATION AND SETTING AGGRAVATION: ASPECTS OF PLACE DISSONANCE

A major purpose of obtaining data on both environmental cognition and environmental preference, it will be recalled, was to assess the extent to which the current environment provided a good fit with the desired residential environ-

TABLE 5.3. Index of Desirability of Different Environmental Elements[a]

Element	Index
Drugstore Food market	.91
Library	.89
Street lights	.88
Bus stop Walkways and pedestrian crossings	.86
Gas station Neighborhood park Post office	.85
Fire station	.81
Specialty food store	.79
A court to play games	.78
Bank or savings and loan Doctor's or dentist's office	.77
Dry cleaners	.76
Beauty and barber shop Elementary school	.75
Specialty store Church or synagogue Children's playground and tot lot Junior high school Friend's place	.72
Restaurant or café	.70
Senior high school	.69
Private or public swimming pool	.68
Clothing or shoe store Hardware store	.66
Hospital or clinic Undisturbed natural/wooded area	.65
Shoe repair shop	.63
Community center Police station	.62
Adult evening classes	.59
Movie theater	.58
Laundromat Appliance repair shop Bicycle path	.57
"Quick-stop" food store	.55
Auto repair shop	.53
Professional offices	.52
Department store Shopping street On-street parking	.51
Day-care center	.49
Outdoor bench on street	.48
Tailor or dressmaking shop	.46
Furniture or appliance store	.45
Gymnasium or health spa Large city park Place of work	.44
Animal care facility	.39
Liquor store	.35
Art gallery College or university	.34
Botanical gardens	.33
Theater for live performance	.32
Museum Relative's place	.30
Antique shop	.26
Freeway access ramp	.19
Business or trade school	.18
Glub or lodge	.17
Skating rink	.14
Bowling alley or pool hall	.13
Beach	.11
Concert hall or opera house	.10
Thrift shop or secondhand store	.09
Bus terminal	−.05
Bar or cocktail lounge	−.07
Amusement park or fairgrounds Sports arena or stadium	−.13
Marina or boat dock	−.17
Alleys	−.19
Zoo	−.20
Discotheque or night club	−.25
Utility lines	−.44
Billboards	−.64

[a]High positive scores mean most commonly desired by all groups; high negative scores mean most commonly unwanted by all population groups.

ment. To the extent that the environment matched expectations, it was consonant with one's desires. To the extent that it did not, it resulted in dissonance. Here, we focus on two essential aspects of place dissonance: setting deprivation and

FIGURE 5.11. An example of setting deprivation—Venice (lower-income white).

FIGURE 5.12. Another example of setting deprivation, settings that were not adequate or did not work—Boyle Heights (lower-income Hispanic).

setting aggravation, which are likely to effect the quality of residential experience. The first type of place dissonance (the mismatch between the current environment and one's expectations) arises when people want certain elements—stores, parks, police stations, and so on—in their area but do not have them. We call this situation *setting deprivation,* after Spivack (1973), who attributed this situation to the physical or psychological inaccessibility of functional places, to their relative scarcity, or to their privileged use. Thus, *setting deprivation* is the sum of the deficiencies in the environment that limit the full range of opportunities for the residents (see Figures 5.11 and 5.12).

A second type of place dissonance occurs when people have environmental settings and hardware in their environment—bars, junkyards, machine shops, and so on—that they do not want. We call this situation *setting aggravation.* It occurs when the presence of certain elements in the environment is considered offensive, objectionable, or irritating (see Figures 5.13–5.19). Operationally,

FIGURE 5.13. Billboards were a common example of setting aggravation—Venice (lower-income white).

FIGURE 5.14. Alleys and utility poles also were not wanted, additional examples of setting aggravation—Venice (lower-income white).

FIGURE 5.15. Nearby refineries were mentioned as aggravating settings—Carson (middle-income black).

FIGURE 5.16. Some examples of aggravating settings—Venice (lower-income white).

we can assess the extent of place dissonance from the measures of neighborhood cognition and preferences for an ideal setting. It will be recalled that three different responses were possible in the card-sorting tasks for both the existing and the desired residential environment. Combining the alternative responses from these two questions established a three-by-three contingency table (Table 5.4). In this table, five of the cells display neutral affects, two display environmental congruence with a positive affect, and two (setting deprivation and setting aggravation) show a negative affect or a mismatch between the desired environment and the present one.

In Table 5.5 we show intergroup differences in these aspects of place consonance and dissonance. It seems that of the seventy-eight settings and hardware considered by our respondents, nearly half were considered congruent with their expectations. Settings or hardware that they felt deprived of were smaller in number. Even fewer were the settings considered aggravating. This finding can

FIGURE 5.17. Further examples of setting aggravation—Bell Gardens (lower-income white).

FIGURE 5.18. More examples of setting aggravation—Carson (middle-income black).

FIGURE 5.19. More examples of setting aggravation—Temple City (middle-income white).

be read as an indication that the current residential areas generally offered more "fits" than "misfits" with our culturally based expectations of a good living environment. But this result need not necessarily be taken as a tribute to the market mechanism or to the public policies that shaped these residential areas. As we will see shortly, the "misfits," although smaller in number, may have played a major role in shaping the quality of the residential experience.

Intergroup differences, according to income class, should be apparent in Table 5.5. The numbers echo once again the principal theme that we have presented: that the degree of place consonance declined (and place dissonance increased) as income declined. However, this was true only in the case of setting deprivation. No significant difference between groups could be seen in the case of setting aggravation. The differences among the three stages in the family cycle

TABLE 5.4. Derived Concepts of Consonance and Dissonance between the Existing and the Ideal Configurations of Environmental Elements

		Description of the existing setting		
		Things actually in in my area	Things actually out of my area	Don't know
Description of the ideal setting	Things I want in my area	Congruent (positive affect)	*Setting deprivation (negative affect)*	Neutral (no affect)
	Things I want out of my area	*Setting aggravation (negative affect)*	Congruent (positive affect)	Neutral (no effect)
	Either in or out is fine	Neutral (no affect)	Neutral (no affect)	Neutral (no affect)

for all three aspects of place consonance and dissonance were comparatively small and statistically not significant.[8]

In Table 5.6, we list the environmental elements that were desired but were not reported as existing in the area by at least 50 percent of the respondents of a particular population group.[9] These lists provide the particulars of our earlier conclusion that the degree of "setting deprivation" was clearly related to income. Whereas the upper-income whites showed no setting deprivation in their area, there was a long list of missing elements for the low-income black and Hispanic respondents. Indeed, one cannot help but notice that, within each income group, the list for nonwhites was somewhat longer than for their white counterparts, suggesting a real or perceived deprivation for reasons of minority status.

Although our data clearly show lower income to be directly related to increased deprivation, what is surprising here is the nature of the elements of which the lower-income groups felt deprived. They were, by and large, not neighborhood facilities at all. Museums, art galleries, department stores, and so on, of which the lower income groups felt deprived, all have traditionally been considered region serving, not neighborhood-oriented, and in an objective sense, none of the residential areas we studied could be said to "possess" these facilities. Our interpretation is that these responses reflected the lack of regionwide mobility of these groups; that is, they saw themselves as being too isolated from regional amenities. Although these elements may be inaccessible to lower-income groups because of sheer distance, cost, the lack of time to partake of them, or even the psychological barriers that had to be overcome to enjoy them, we believe that these lists are also symbolic of the general deprivation the lower-income groups' experience.

TABLE 5.5. Aspects of Place Consonance and Dissonance by Population Groups and Stages in the Family Cycle[a]

	Means for population groups						
Aspects of place consonance and dissonance	Upper white (n = 85)	Middle white (n = 80)	Middle Hispanic (n = 59)	Middle black (n = 86)	Lower white (n = 88)	Lower Hispanic (n = 55)	Lower black (n = 22)
Number of congruent settings	48	45	46	43	38	36	39
Number of deprived settings	7	8	10	10	12	14	24
Number of aggravating settings	3	4	4	4	4	4	3

	Means for stages in the family cycle		
	Households with children (n = 256)	Households without children (n = 117)	Elderly (n = 102)
Number of congruent settings	43	44	40
Number of deprived settings	11	10	10
Number of aggravating settings	3	5	3

[a]Adjusted means from a multiple classification analysis involving a two-way analysis of variance, in which population groups and stages in family cycle factors were introduced simultaneously, showed little change from these numbers. Hence, the unadjusted mean scores are shown here.

TABLE 5.6. Setting Deprivation and Setting Aggravation by Population Group

Upper-income white ($n = 85$)	Middle-income white ($n = 80$)	Middle-income Hispanic ($n = 59$)	Middle-income black ($n = 86$)	Lower-income white ($n = 88$)	Lower-income Hispanic ($n = 55$)	Lower-income black ($n = 22$)
			Setting deprivation (Want it but don't have it)			
	Natural/wooded area	Natural/wooded area	Place of work[a]	Natural/wooded area	Botanical gardens	Antique shop
	Bicycle path[a]	Art gallery[a]	Art gallery[a]	Museum[a]	Animal care facility[a]	Department store
	Open space[a]	Museum[a]	Theater for live perf.[a]		Department store[a]	Shopping street/mall
		Botanical garden[a]	Bicycle path[a]		Shopping street/mall[a]	Professional offices
		Amusement park[a]	Movie theater[a]		Art gallery[a]	Business or trade school
		Bicycle path[a]	Skating rink[a]		Museum[a]	College or university
					Theater for live perf.[a]	Place of work
					Concert hall/opera[a]	Art gallery
					Amusement park[a]	Museum
					Bicycle path[a]	Theater for live perf.
					Large city park[a]	Club or lodge
					Skating rink[a]	Bicycle path
					Bus terminal[a]	Bowling alley
					Open space[a]	Movie theater
					Natural/wooded area[a]	Skating rink
						Open space
						Natural/wooded area
						Animal care facility[a]
						Appliance repair store[a]

						Clothing or shoe store[a] Furniture/appliance store[a] Hardware store[a] Restaurant/cafe[a] Botanical gardens[a] Concert hall/opera[a] Zoo[a] Amusement park[a] A court to play[a] Gymnasium/spa[a] Large city park[a] Marina or dock[a] Children's playground[a] Sports arena/stadium[a] Bus terminal[a] Freeway ramp[a] Billboards[a]
Setting aggravation (Have it but don't want it)						
Utility lines[a]	Utility lines Billboards	Utility lines[a] Alleys[a] Billboards[a]	Utility lines[a] Billboards[a]	Utility lines[a] Bars/cocktail lounges Billboards	Alleys Liquor store[a] Bars/cocktail lounges[a] Billboards[a]	Liquor store Alleys Bars/cocktail lounges Gymnasium

[a]Environmental elements identified by at least a third but not more than a half of the respondents. The rest of the items were identified by a half or more of the respondents.

TABLE 5.7. Measures of Association (Kendall's Tau C) between Overall Evaluation of the Residential Area and Three Measures of Place Consonance and Dissonance[a]

	First five groups (n = 398)	All groups (n = 475)
Total number of "congruent" settings	−.21[b]	−.21[b]
Total number of "deprived" settings	.22[b]	.22[b]
Total number of "aggravating" settings	.03	.02

[a]Evaluation scale: 1 = excellent; 2 = good; 3 = average; 4 = fair; 5 = poor. The negative measure of association, thus, indicates that the higher the number of congruent settings, the better the evaluation of the residential area. Similarly, a positive measure of association indicates that the higher the number of deprived or aggravating settings, the worse the evaluation categories of five-point intervals.

[b]Significant at the .001 level.

We suggested earlier that the degree of match or mismatch between the desired and the existing set of environmental settings and hardware can explain overall residential satisfaction and well-being. This argument is implicit in Spivack's (1973) theory of "archetypal place," and it is the concept of "mental congruence" suggested by Michelson (1968). In developing a general theory of satisfaction with realms of life experience, Campbell *et al.* (1976) also emphasized the role of cognitive standards in determining satisfaction with different life experiences. Our own findings support this view. Table 5.7 clearly shows the nature of the association between place consonance and dissonance and the evaluation of residential environment. Because Kendalls' tau is a measure of the strength of association between variables, the figures shown here suggest that feelings of deprivation may play a more significant role than feelings of aggravation in residential area evaluation.

DISTANCE THRESHOLDS OF DESIRED SETTINGS AND HARDWARE

Finally, we turn to another operational question that most physical planners and designers are likely to ask: How close to the home should these settings and hardware be located? What is the threshold distance for satisfaction? Although we did not ask our respondents directly about the acceptable distance of these desired elements, we can estimate indirectly some approximate measures for desired distances. These measures are based on responses to questions about the time it takes to reach the farthest edge of one's residential area.[10] We have used this information in the following way: If an individual indicated that it took x minutes to reach the farthest edge of her or his residential area, then we assumed that all settings and hardware that she or he wanted "in" the residential area, ideally, should be located at a distance no greater than x. In aggregating these individual responses, we chose to use the "median" as the relevant measure because it gives the planner some sense of threshold, that is, the maximum that can be allowed to make at least 50 percent of the residents satisfied. Because the respondents could indicate the distance measure (in minutes) in terms of different travel modes—walk, bike, drive, motorcycle, bus—the distance threshold is shown in two different diagrams (Figures 5.20 and 5.21) for walking and driving, which were the two predominant modes chosen. Thus, it could be proposed that the majority of people who would typically drive their cars to do their shopping, for instance, would find a drugstore or a food market in less than nine minutes' drive if they were upper-income whites, in less than ten minutes if they were middle-income whites or lower-income Hispanics, in less than six minutes if they were middle-income Hispanics, in less than five minutes if they were low-income whites, and so on. In Figures 5.20 and 5.21, we show such distance

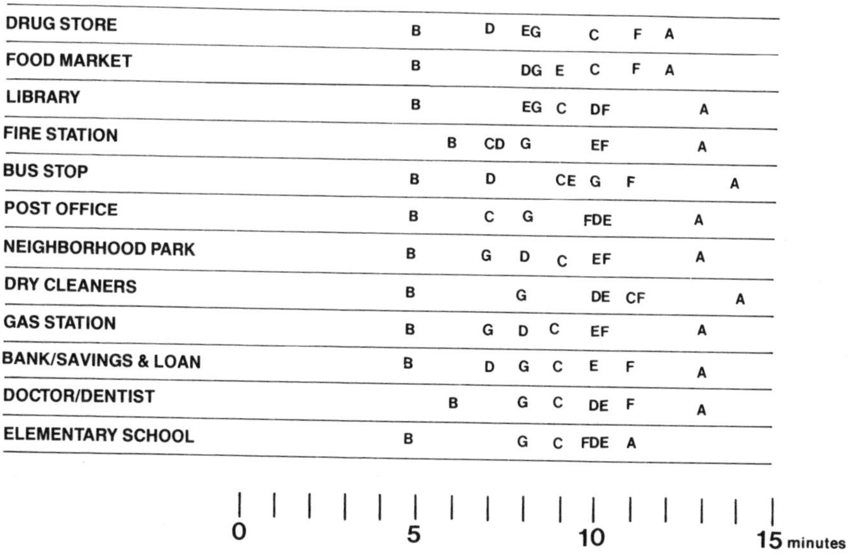

Setting	Groups (in order along the 0–15 minute scale)
DRUG STORE	B D EG C F A
FOOD MARKET	B DG E C F A
LIBRARY	B EG C DF A
FIRE STATION	B CD G EF A
BUS STOP	B D CE G F A
POST OFFICE	B C G FDE A
NEIGHBORHOOD PARK	B G D C EF A
DRY CLEANERS	B G DE CF A
GAS STATION	B G D C EF A
BANK/SAVINGS & LOAN	B D G C E F A
DOCTOR/DENTIST	B G C DE F A
ELEMENTARY SCHOOL	B G C FDE A

0 5 10 15 minutes

FIGURE 5.20. Median driving distance (in minutes) to the most commonly desired settings by population groups. A = upper-income white; B = middle-income white; C = middle-income Hispanic; D = middle-income black; E = lower-income white; F = lower-income Hispanic; G = lower-income black.

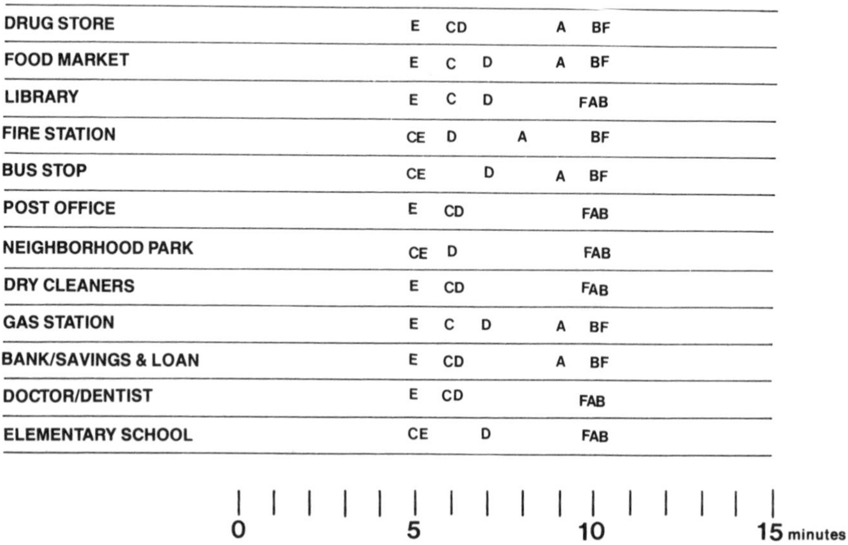

Setting	Groups (along walking-time scale)
DRUG STORE	E CD A BF
FOOD MARKET	E C D A BF
LIBRARY	E C D FAB
FIRE STATION	CE D A BF
BUS STOP	CE D A BF
POST OFFICE	E CD FAB
NEIGHBORHOOD PARK	CE D FAB
DRY CLEANERS	E CD FAB
GAS STATION	E C D A BF
BANK/SAVINGS & LOAN	E CD A BF
DOCTOR/DENTIST	E CD FAB
ELEMENTARY SCHOOL	CE D FAB

0 5 10 15 minutes

FIGURE 5.21. Median walking distance (in minutes) to the most commonly desired settings by population groups. A = upper-income white; B = middle-income white; C = middle-income Hispanic; D = middle-income black; E = lower-income white; F = lower-income Hispanic; G = lower-income black.

thresholds for the twelve[11] elements that were most strongly desired, with some degree of consensus, by almost all income groups and stages in the family cycle.

It will be noted that, although within a particular group variations of the distance threshold for different settings and activities were minor, and in some cases nonexistent, the differences between groups were quite significant.[12] However, unlike the size of the residential area maps, there was no clear pattern in the distance thresholds following income differences.

There was, however, a substantial difference in the physical distance implied by the thresholds for walking time versus driving time. For example, for the middle-income whites, a five-minute walk was the threshold distance for a food market or a drugstore. Assuming a three-mile-per-hour pedestrian walking speed, a five-minute walk translates into a distance of two average city blocks. But for the same population group, the majority of drivers would have found a drive of less than ten minutes acceptable. Assuming an average speed of 20 miles per hour, that distance translates into about 25 average street blocks. For the planner engaged in the operational aspect of residential planning, this need not pose an anomaly. We do not specify these measures as precise standards; they are merely guides. Planners may choose their own decision rules, based on density, location, inner-city versus suburban sites, and assumptions about the lifestyle, the life cycle, and the driving habits of future housing consumers. Alternatively, planners may use the lesser of the two thresholds (walking time) and be confident that the distance will be satisfactory to most drivers as well.

Finally, this discussion would be incomplete if we failed to consider how these distance thresholds compare with those specified in *Planning the Neighborhood.* In every case where there is a comparable distance standard available, we find that our data correspond to and are in some cases well within those previously formulated standards, as shown in Table 3.6. For the most part, this correspondence suggests a vindication of the standards in *Planning the Neighborhood.* In a few cases, however, it is obvious that the earlier standards would today leave the majority of certain social groups somewhat unhappy.[13]

SUMMARY AND CONCLUSIONS

In this chapter, we have attempted to focus on the physical composition of the residential area. On the assumption that a good residential area is defined not just by the social context or the housing, but also by the various environmental settings and hardware that are desirable and necessary in day-to-day activities, and that are a part of the quality of residential life, we have tried to identify here the settings and hardware desirable for such purposes. In the process of identifying these by population groups and stages in the family cycle, we have also developed a scale of the residential desirability of settings and hardware that can suggest the psychological distance between various settings and one's home. The checklist of residential settings, generally—and such a scale, specifically—can be used to evaluate existing residential environments and to establish targets and priorities for improvement.

To establish the importance of these place characteristics in influencing one's perception of one's residential area and overall sense of well-being, we explored the concept of *place consonance,* that is, the degree to which a residential area provides a good fit with the expectations of a resident. In this context, we introduced the concepts of *setting deprivation* and *setting aggravation,* along with the concept of setting congruence, as operational measures of place consonance. It was shown that the measures of place congruence and setting deprivation, at least, were significantly associated with overall measures of evaluation of the residential area. There was a further suggestion in the data that the measure of setting deprivation may even be a better predictor of place evaluation than the measure of setting congruence.

Finally, we attempted to develop some operational measures of distance threshold for the desired settings and hardwares for different populations groups. We hoped these would provide a practical guide for the planner and designer currently engaged in developing a subdivision plan, in setting up a residential zoning ordinance, or in proposing changes to improve the conditions of an inner-city neighborhood.

SOME NOTES ON THE DIFFERENCE BETWEEN MARGINAL AND ABSOLUTE PREFERENCES

In light of the clear showing of differences in the quality of the residential environment the social acceptance of neighboring inhabitants, the size of the residential environment, and the number of activities by income groups, it is important to add a confounding note. In our survey, we frequently asked people to evaluate their environment or "holding." Moreover, we asked them to tell us what they would like to have in the way of access, ambient qualities, settings and hardware, and so on. There were clear differences according to income group. But were we getting a "true" reading from our respondents? How much of their response was biased by such things as exposure to desirable residential areas (or the lack thereof), friendly and helpful neighbors, good air quality, and accessibility, as well as education, job opportunities, and so on—all of which might influence the type of response.

We therefore believe that the reader should keep in mind an important distinction that we came to appreciate only after thinking about what these responses meant.For one, we must distinguish between "marginal" and "absolute" preferences. The *absolute* is a measure of the basic priorities or the total degree or amount of some residential attribute necessary to achieve some ideal. *Marginal* refers to the priorities or the amounts necessary in addition to what the respondent already possessed to achieve that ideal. Thus, when people talk about their preferences and priorities, we should know what type they mean.

Also, we should know what determines preferences. If residential preferences are determined solely by differences in lifestyle, social mores, and stage in the family life cycle, then responses to inquiries can be interpreted in a straightforward manner. But if preferences are also based on what one knows about existing residential environments in general (as well as about what one possesses), then we must also know how much of each environmental attribute is currently possessed or known about. If there are current differences in knowledge and "holdings," these differences will cause differences in expressed preferences.

This latter situation raises some perplexing questions with regard to the cultural pluralism argument stated above. If differences in income enable different social groups to command different levels of quality in residential settings, or if social discrimination and segregation work the same result, then any differences in marginal residential preferences as expressed by different groups may be merely the result of income and social inequalities and may not reflect any "true" differences in preferences, all else being equal.

Thus, present residential inequalities become a salient consideration in (1) understanding differences in levels of present satisfaction, future preferences, and priorities for change; (2) developing criteria for future residential environments for different social groups; and (3) addressing the thorny question of how to attain equity in the allocation of resources for different social groups if they currently have different ''holdings'' of residential amenities. Accordingly, we have tried to be fair to the reader by attempting to show ''baseline'' holdings for each group, as well as by revealing their residential impressions, evaluations, and preferences. In this way, we allow the reader to make mental adjustments for the marginal versus absolute biases that are inherent in such responses, but we do not know how to compensate for these potential biases in a precise way. We do, however, address the policy implications of such inequality in a more general way in presenting a normative model of a good residential environment in Chapter 7.

NOTES

1. Myer Spivack (1973), who built on Barker's concept of a ''behavior setting'' to develop a theory of ''archtypal place,'' expresses similar views in reference to the precise and specific definitions given by Barker.
2. After the term *neighborhood hardware,* suggested by Steve Pierce (1976) in his analysis of our data. Although the term *neighborhood hardware* is perhaps more elegant, it suggests to us a risk of being inadvertently associated with the ''neighborhood unit'' and its related principles.
3. With the exception of the elements that can be best described as ''environmental hardware,'' all of the settings correspond to one or more activities presumed to have some bearing on residential satisfaction and well-being (Chapin, 1974).
4. Only the settings and the hardware that were desired by four or more groups are included in this table. Also desired (usually weakly) by three or less groups are, in order of collective desirability: gymnasium or health spa; liquor store; theater for live performance; thrift shop or second hand store; botanical garden; club or lodge; bowling alley or pool hall; amusement parks or fairgrounds; beach; skating rink; antique shop; museum; sports arena or stadium; bus terminal; and alleys.
5. Whether this seemingly higher tolerance for nonresidential land use among the minority groups stemmed from their current experience with environments that abounded in many different land uses, or from a sense of locational isolation from many of these facilities, can only be a matter of speculation. Our data are inadequate to offer a firm answer.
6. Neither did they want to be near a zoo (an understandable desire), but why not near a beach? We do not know whether this reluctance to be near a beach—which is commonly seen as an amenity—reflects a perception of a high cost of living or a dislike of the leisure-seeking crowd and the resulting congestion.
7. Here, we have chosen to use the first five population groups (i.e. excluding the lower-income Hispanic and black respondents) because of the questionable nature of the responses from the last two groups, as discussed elsewhere. In developing this composite index, underrepresented groups were weighted to produce a uniform proportional distribution of stages in the family cycle within population groups of equal size. It is to be noted that proportional distribution by income class (ignoring ethnic differences) in this aggregate sample is, 1 (upper): 3 (middle): 1 (lower); and that of the stages in the family cycle is, 2 (households with children): 1 (households without

children): 1 (elderly). We have assumed that this proportional distribution approximated the demographic profile of the larger metropolitan area. A different weighting scheme would, of course, produce a somewhat different ordering of the elements.

8. The effects of population group, when entered simultaneously with the stage in the family cycle in a two-way analysis of variance framework, were significant at a .001 level for both the setting congruence and the setting deprivation variables, (using first five and all seven groups separately.) The effect of stages in the family cycle was found to be not significant in all instances. No significant interaction effects were noted.
9. When organized by the stages in the family cycle, very few items appear on the list of setting aggravation and deprivation, and most of them were identified by less than half of the respondents in each category. Therefore, we did not include a table here, as it would not add anything significant to our discussion.
10. See questions 9A and 9B, Appendix I.
11. Of the fourteen most strongly desired elements, two—''streetlights'' and ''walkways and pedestrian crossings''—fell into the ''environmental hardware'' category. The notion of distance threshhold was not particularly meaningful for these elements. Hence, these two items are not included in Figures 5.20 and 5.21.
12. However, we have no good explanation for such differences between social groups. We were not able to notice any clear pattern in such intergroup differences.
13. For example, according to the PTN standards, recreation facilities should be within 20 minutes' travel time (see Table 3.6), which is substantially higher than the distance thresholds for recreation-related facilities such as neighborhood parks, as shown in Table 3.6.

6

Taking Stock

A Synthesis of the Findings

It is time to summarize what we have covered. In the last three chapters, we have presented selected findings about how the residents of twenty-two Los Angeles neighborhoods perceived their residential environments and about their residential needs, images, values, and priorities. We have examined the concept of residential environment within three different frames of reference: In Chapter 3, we tried to capture the essence of the residential area in the broadest sense, so as to include all of the different meanings and functions it may have for different individuals. Here, our focus was the residential *milieu*: the social, the physical, the functional, and the symbolic. In Chapter 4, we emphasized the *form* of the residential area. Here we explored the spatial and territorial dimensions of the residential area implicit in our respondents' cognitive maps. Finally, in Chapter 5, we examined the residential area as a *setting* for daily activities. Here, we identified the environmental settings and hardware that belong to the residential area and those that should be excluded from it.

Each of these perspectives carries with it implications for public policy. The definition and descriptions of the residential milieu have provided the various sets of concerns and priorities held by different resident groups. These help us to determine the domains of public policy through which improvements in residential quality of life can be sought. The images of residential areas, on the other hand, give important clues to salient features of residential space. The information on the environmental settings and hardware desired (or not) by people in their residential areas provide an operational basis for determining standards for acceptable land use, public facilities, and location.

Before we discuss specific proposals for future residential planning and design, it is worth reviewing the salient points of our findings, not just for the sake of a global summary, but for the purpose of linking the three levels of perceptions—milieu, form, and setting—in a coherent fashion. This is also the place to examine the fundamental differences in these perceptions by income, ethnicity, and stages in the family cycle—the three main dimensions in which we categorized our respondents. Thus, the following discussion synthesizes and interprets what it all means to us and sets the stage for the specific proposals and recommendations contained in Chapter 7.

A SUMMARY OF THE FINDINGS AND THEIR IMPLICATIONS

The Residential Area as a Milieu

In Chapter 3, we attempted to analyze our respondents' impressions of their residential areas. We used the term *residential area* in framing the initial questions, so as not to influence our respondents' judgment about whether they lived in a "neighborhood" (with the attendant connotations and denotations), and we used a nondirective, open-ended approach to questions so as not to predetermine the nature or dimensions of the responses. Indeed, we hoped that we would uncover alternative constructs of the residential area, if they existed, so as to judge better the merits of the neighborhood unit concept and to determine if alternatives should be devised for the environmental design professions.

Despite the physical design orientation of our purpose, we found, generally, that social classifications, not physical design, were the predominant means of organizing respondents' concepts, and that social class or "the kind of neighbors living nearby" was the descriptor that most readily came to their minds. We also found that physical design played an important role in the residential environment, as well as in producing significant effects, albeit not the role that its creators had assumed, nor the effects that its practitioners intended. Finally, we uncovered some other dimensions pertinent to thinking about residential areas that had not been included in the original design concept.

Our upper- and middle-income respondents, in general, used favorable adjectives to describe their neighbors. But the lower-income groups were more apt to mention the *variety* of neighbors, many of whom they did not like or trust and, indeed, whom they decidedly feared at times. In fact, the notion of *neighbor* largely disappeared at this level and was replaced by a sense of merely "other people" who also lived in the area.

The mention of fear raises another point salient to all our respondents: personal and property safety. The degree of "safety" present in their area was a concern of all income groups, but the concern varied by income class. Upper-income groups mentioned property safety (robberies) in this context, whereas middle- and lower-income groups increasingly stressed the added concern of personal safety (muggings). Few respondents, however, related these concerns to the physical features or the design aspects of the residential area.

Although the social milieu and personal/property safety were the most important ambient characteristics, air quality was also of major concern. The presence or absence of smog or dirty air was mentioned frequently by all groups of respondents. Predictably, its source (or the lack of it) was attributed to natural conditions specific to the location of the residential area, not to local design features.

The amenities and conveniences of the residential area, or the lack thereof, were also mentioned. Generally, people seemed to have done better in convenience than in amenities. Only the rich boasted of a high level of amenities in their environment. The middle-income groups had fewer environmental amenities, but they considered themselves at least conveniently located with respect to most public and private facilities. Unfortunately, the low-income groups, particularly the minority groups, could not even claim such locational advantages.

The sense of the physical place was a recurring theme but was apparently of less importance. Here, people mentioned the "look" and "feel" of their residential area. Once again, the upper-income groups typically took pride in the appearance and the atmosphere of their areas, praising the landscaping, the area's natural beauty, its physical layout, and the exclusive land-use. The middle- and lower-income people spoke mostly about the absence of pleasant sensory qualities and the absence of a sense of place. They were typically concerned about noise, traffic, dirt, litter, unpaved streets, or bad street lighting. The physical place appeared to be an important component of community identity for the rich, but the other groups did not make such claims.

Interestingly, the *location* of public improvements—which is a prime focus of the neighborhood unit construct—received less mention than did the quality of the service associated with the improvement. Thus, investment decisions about where to locate a facility or how to incorporate it into the physical design of the area appeared to be less important in the respondents' minds than were the annual operating decisions or the quality of service provided by the facility once it was installed, regardless of where it was located.

Overall, these concerns about the residential milieu can be translated into a common set of characteristics for all groups, as is evident in the data on residential satisfaction and priorities. That is, if the residential environment was seen as a "bundle" of attributes—some pertaining to the *ambient* qualities and others representing *access* opportunities—it was the ambient attributes that typically ranked higher than the access attributes (Banerjee *et al.*, 1974). Not unexpectedly, most of our respondents were more dissatisfied with the ambient qualities of the environment, and the priorities may indeed reflect an inverse relationship with the current satisfaction. But the rich also valued ambience over access, even though they were equally satisfied with both. Is it possible that these priority orderings are independent of present satisfaction levels?

What do these findings mean for the neighborhood concept and *Planning The Neighborhood* both of which are premised on beliefs in physical determinism, and both of which were designed to create a congenial and supportive social milieu? On the surface, the design of the physical environment would seem to be relatively unimportant in any direct sense. Certainly, this is what many sociologists have told us for quite some time (Gans, 1968; Keller, 1968; Webber, 1963, 1964).[1] Nevertheless, we would like to suggest here that, in an indirect sense, design can still be important, and that designers' instincts have often been correct, even though their emphases have frequently been misplaced.

For one thing, the design *per se* is not as important in creating congenial and supportive social milieus as is the *cost* of the design, for the cost of transforming a design concept into three-dimensional physical form is extremely influential in determining who can afford to live in an area, and hence who one's neighbors will be. *In short, design determines cost; cost determines inhabitants; and inhabitants determine residential satisfaction.* We return to this theme at the end of the chapter.

Second, although crime and safety issues were not the cornerstones of the principles of neighborhood unit, they are related to design concepts. The link between crime and the physical environment has only recently been formally addressed (Gold, 1970; Ward, 1973). Newman's (1972) argument that site planning and architectural design can deter or facilitate criminal activities by creating a "defensible space" is of interest to policymakers.[2] While our respondents were more concerned about the people who commit the crimes, future models of residential design and planning should give attention to those features of the physical environment that can influence the incidence of crime (e.g., the presence of unlighted and unwatched alleys).

A third aspect of design and form is the location of public and private facilities. Here, our respondents often suggested emphases different from those that had been intended in the original concept. The success of the neighborhood design construct was particularly evident with respect to the location of schools, where the neighborhood unit principles adopted by the school district planners have clearly resulted in a successful pattern, as evidenced by the high satisfaction with access to schools in most locations and among most social groups. On the other hand, the convenience of shopping, so frequently mentioned by middle-income and some low-income residents, may be more of a tribute to private market forces than to public planning, for shopping often takes place at shopping centers of a magnitude never envisioned by the originators of the neighborhood unit, who thought largely in terms of "ma and pa" convenience stores.[3] Here again, income enters the picture in terms of location based on "effective demand.'' Suffice it to say here that the placement of shopping centers is frequently a function of a regional perspective, not the neighborhood orientation.

Moreover, the emphasis on the location and the type of facility within the neighborhood unit neglects what our respondents reported to be a more important facet of residential life: the *quality* of the services at the location. Our respondents were frequently willing to give up easy accessibility for improved quality in service, thus making immediate proximity of the facility of less importance than had been ascribed to it by the original concept. Nor did the concept encompass the full range of residential services that our respondents deemed important in their own frame of reference. Police and fire protection, garbage collection, street maintenance, and so on were included in our respondents' notion of the quality of their residential area and their evaluation of it.

Finally, the complexity of urban life suggested that an implicit metropolitan perspective was necessary as well as a local neighborhood one. We have already mentioned the regional perspectives of shopping-center developers and tenants (and their apparent success, for the most part, as measured by respondent satisfaction); air quality is another instance. Although air quality was an important concern to the respondents, the problem cannot be successfully addressed solely from a neighborhood design perspective. To be sure, recent studies indicate that local air quality can be improved somewhat through managing traffic and creating auto-free zones (Thullier, 1978), but the greatest improvements can be accomplished only at the regional air-shed level, beyond the purview of the residential area designer. Similarly, the problems of crime and public safety are better addressed at the city level rather than at the neighborhood level, where design acts to deter crime but does not deal with the fundamental social causes of crime.

Indeed, the ordering of priorities between the access and the ambience components of the residential bundle of attributes clearly points out that those things that matter most can be affected largely by metropolitanwide policies, rather than by local improvements as envisaged by the neighborhood unit concept. It suggests that the physical form and the social ecology of the larger metropolitan community may have a lot more to do with the residential experience at the local level than do local improvements themselves. Although an accessible and redundant physical form increases choice and maximizes convenience, it could also contribute to an uneven distribution of ambient resources. This is what the satisfaction and priorities of our respondents suggest about the metropolitan form of Los Angeles. It may be increasingly important to focus on the form and the organization of the metropolitan community, as well as on the residential space encompassed in the neighborhood unit concept. If social equity and distributive justice are to be pursued in residential planning and design, the search for a new paradigm must begin with a sense of the overall structure and organization of the larger community space.

The Form

The above findings were elicited through verbal responses. Although these kinds of responses can describe and evaluate, they are less capable of describing the relationship between objects in space. This description can be better accomplished through maps or diagrams. In Chapter 4, we presented maps that our respondents had drawn of their perceived residential area. The results ranged from the schematic and abstract to the detailed and pictorial. The images portrayed ranged from a mere street intersection to a block face, to an entire block, and even to a collection of blocks. The territory shown varied from a loose collection of points, places, and landmarks strung together, with little sense of boundary, to a tightly integrated and well-demarcated area. Collectively, these maps suggested that the salient features found in the neighborhood unit concept were important to only a few. Clearly, only a few people conceived of place and location in a residential area along the highly rationalized and hierarchical lines so carefully drawn up and explained by the environmental designers who subscribe to the neighborhood unit concept.

Moreover, to the extent that there was an organizing schema to these maps, it belonged to the city street and the major thoroughfare grid or network, rather than to the cells in between. Whereas the neighborhood concept deplored the ravages and accoutrements of the automobile—noise, exhaust, and a voracious appetite for right-of-way and parking (not to mention dangers to pedestrians)—our respondents indicated that these are mixed blessings not necessarily to be decried.

Whereas the neighborhood concept looked inward to a single core of interest, our respondents showed multiple nodes on their maps. Whereas the neighborhood concept used major streets as a boundary to exclude others, our respondents drew nonbounded, open-street systems that connected with adjacent parts of the city. Whereas the neighborhood concept attempted to keep the car at bay and to confine it largely to a peripheral major street, which was itself obnoxious and destructive to family life, our respondents focused to a considerable extent on the importance of the "evils" in serving as major communication links in the city system. The maps appeared to emphasize the *urban* interaction achieved *along* these routes, rather than the *village* interaction between neighbors, taking place *within* the residential blocks, as emphasized by the proponents of the neighborhood unit.

There would appear to be a paradox here. The verbal responses reported in Chapter 3 emphasize the predominance of social definitions of a residential area that would presume interaction between residents; hence, we would have anticipated the frequent sketching of residences on the maps. Yet, the pictorial responses that emphasized the street system suggested lines of communication *along* streets more than *between* residences. Which is really more important? Or did the two different methods of response merely allow the respondents to highlight important but different facets of residential life? We believe that the maps denoted the important physical objects (in contrast to the social interaction) suggested by the verbal responses. Furthermore, contrary to the neighborhood unit's emphasis on residences, community centers, shopping center, schools, and so on, here the streets were the most salient physical object, perhaps because they were the primary physical means by which links were maintained between these objects. That is, it is the street *system* which allows people to connect with a variety of destinations, that is seen as important, not the destinations themselves. This might not be so surprising if one were to chart the frequency of use. The street is common to every trip; the store, the school, and the doctor's office are important only on the trip that has that destination specifically in mind.

Although the street system was a predominant landmark for all groups, there were differences between groups as well. For example, the upper-income groups, as was the case in their verbal descriptions, were the most detailed in their depiction of their residential area. They showed a comparative opulence of environmental amenities and resources, in much the way that environmental designers envision a well-stocked residential area. The middle- and lower-income groups, on the other hand, drew attention to commercial establishments as salient aspects of their residential area. Indeed, franchise names and other corporate symbols played a significant role in the landscape in these residential areas, although such features are normally beyond the purview—and even the ken—of the neighborhood unit concept. The importance of commercial establishments was further emphasized by the landmark quality of major shopping centers, whose service area encompassed an area that would contain several traditional neighborhood units, and whose salience as important nodes was identified by all income groups.

Although public facilities were noted by our respondents, they were usually considered less important landmarks, despite the importance assigned to them in the neighborhood unit concept.

Thus, we note an important omission on the part of the neighborhood concept, and indeed an omission that is common among many environmental designers as well. Typically, design standards concentrate on public facilities as deserving prime attention (although adequate commercial parking is also viewed as an area of concern for designers), whereas our respondents focused far more on private facilities, including corporate or brand names. The mention of brand names (e.g., McDonald's or Thrifty Drug) may indicate an environmental and residential quality of importance far greater than designers have recognized. Environmental designers may acknowledge the need for fast-food chains or drugstores, but many do not appreciate quality differences as well. The private sector spends enormous sums to help consumers perceive differences between, say, McDonald's and Burger King, Thrifty Drug and Rexall. Whereas parks may not differ much in quality from one community to another (the designer merely calls for a community park in the plan, not for a brand name of park), there may be a substantial difference between supermarkets in the eye of the resident/consumer.

The size of the residential area shown on the maps also varied with income and, at times, with race/ethnicity. The upper-income groups drew areas that frequently were larger than the half-mile diameter suggested by the neighborhood unit. The low-income groups, especially the minorities, depicted an area substantially smaller than the area recommended in the neighborhood unit. Furthermore, neighborhood scale was not the sole organizing spatial concept; sometimes, the area depicted encompassed many neighborhoods and one or two regional shopping centers. Even a business area was used to orient the residential area and the spatial cognitive map.

These discrepancies between the neighborhood unit concept, on the one hand, and the variety of concerns and images reported by our respondents, on the other—some at considerable odds with the neighborhood unit concept—raise once again the basic issue of the validity of using a neighborhood concept as an organizing device in the first place. Does the "neighborhood" really exist? To be sure, part of the differences reported might lie in our alternative use of the term *residential area* in the early questions in the interview.

To test this possibility without boring our respondents with repetition, we simply asked them whether there was a difference in size between their residential area and their neighborhood and, if so, to draw their neighborhood on the same map. The results of this question are not easily interpreted, for 40% said that they were the same size, 40% said that the neighborhood was smaller, and 20% said that it was larger. At best, we can suggest that there is frequently a two-tiered concept in people's minds, although it may not be formed with great clarity.

Nevertheless, despite some confusion (or at least lack of agreement) in people's minds about size, a slight majority considered it important to live in a neighborhood; a much larger number considered living in a neighborhood of at least some importance. But when we examined these outcomes by race/ethnicity, we found a surprising result: although the neighborhood concept was supposed to have been designed for middle-income whites with children (the needs and preferences of other ethnic/racial groups were not really considered in its formation), minorities, in general, considered neighborhood living more important than did whites. The stage in the family cycle was also important here: parents with children were much more apt to think living in a neighborhood was important than were those households of the same general age without children; yet, interestingly, the elderly were even more likely to want to live in a neighborhood than either of the other two groups.

In general, the reasons given for the importance of living in a neighborhood were generally similar to those features initially volunteered by people as characterizing their residential area. That is, social reasons (sociability, friendliness, family-related) predominated; although physical and environmental amenities, convenience and shopping also were viewed as important.

The Physical Setting of the Residential Area

The neighborhood unit concept did have the merit of discussing only those considerations about which the designer could do something in a very direct sense. Mindful of this fact, and aware that, for all the limitations of physical design, the location and placement of sites and facilities in the residential area were not entirely unimportant, and assuming that the environmental designer had control over these decisions, we specifically examined the residential environment as an area defined by physical settings and hardware. Here, we assumed that the presence or absence of what we called *environmental elements* (e.g., gas stations, billboards, bus stops, and dry cleaners) could definitely contribute to residential satisfaction and well-being, and we wanted to learn which of these elements were desirable within a residential area and which were better located outside the area. We explored these issues in Chapter 5.

Here, we detected a core of similar likes and dislikes among the groups, and a decided divergence beyond this core. The number of elements that we treated (78 in all) makes a brief summary difficult here because our purpose was to provide specific checklists by income, race/ethnicity, and stage in the family cycle for the designer to consider, as well as some indication of the degree of strength with which these preferences for different elements were held.

Next, we examined the extent to which our respondents' own residential areas conformed to their desires. Here, we analyzed the degree of *setting deprivation* (the extent to which the things that people wanted in their residential area were missing from them) and *setting aggravation* (the extent to which things were present in their area that they wished were somewhere else). As one would expect, setting deprivation increased as income decreased; but, surprisingly, there was no significant difference between groups in terms of setting aggravation. The stage in the family cycle did not appear to be significant in either case.

But merely knowing whether an element is desired within a residential setting is not specific enough for good planning. The desired accessibility of or proximity to that item is also a necessary bit of information. Thus, we provided a list of maximum distances for the most commonly desired elements by population group. Accessibility or proximity was measured in time rather than distance, so as to allow for alternative modes of transport and also for a comparison of alternative design solutions. We also compared these findings with the recommended standards in PTN and found, with few exceptions, that most of the access standards were still adequate.

POPULATION GROUP DIFFERENCES IN EXPERIENCING THE RESIDENTIAL ENVIRONMENT

To this point, we have summarized our findings on the milieu, the form, and the setting of the residential area, and we have compared these with the premises and components of the neighborhood concept. In doing so, we have at times touched on differences exhibited or expressed by different population groups. But one of our purposes in the research was to explore very specifically whether and where these differences existed, and whether they differed enough to have policy significance over and above statistical significance. We wanted to know if alternative design paradigms were warranted to accommodate these differences?

The most pronounced differences are due to income. This finding has been supported in numerous ways: in the verbal responses to our open-ended questions; in the tables of descriptive data; in the maps that our respondents drew; and, where it has been possible to apply a two-way analysis of variance to examine adjusted means (after controlling for the effects of the stage in the family cycle), in different measures of residential experience. Moreover, it showed strongly in our activity analysis, which we do not report here. We have extreme confidence in this finding, for it has been supported from so many different approaches and avenues of investigation (Chapin, 1974).

It is not surprising that income was the most significant variable for our groups; but what is baffling is that it was so much more pronounced as a finding than either ethnicity or stage in the family cycle. As we will indicate, it both simplifies the problem of providing modifications to PTN or devising alternative paradigms and starkly reveals an underlying social reality of inequality that cannot be dealt with effectively by a simple variety of design offerings.

Although some difference in the preferences of ethnic groups (holding income class constant) can be seen in our data, it is not clear whether such differences reflect culturally rooted value differences (as the cultural relativists would like to argue) or whether they are a result of existing environmental conditions. For example, the list of settings and hardware desired in the residential area was longer for nonwhites than for whites of both low and middle income. Does this finding reflect an ethnically based value difference in relative desires, or is it a reflection of a higher degree of chronic setting deprivation among these racial groups? The same question can be raised in the case of intergroup variations in the perceived importance of neighborhood living, for which the minority groups showed a greater preference. Do they have stronger preferences for some "neighborhood ideal" because, in reality, their residential areas approximate this ideal less than do the areas inhabited by whites? There is a strong possibility that differences in preferences between racial groups can be explained by arguing that our elicitation of preferences obtained marginal rather than absolute preferences. In short, what people said they wanted might have been determined by how much (or how little) they had at the time, and what they had was frequently a product of discrimination superimposed on a market-oriented economic system.

If a difference in expressed preferences for environmental conditions is an artifact of social discrimination rather than being based on unencumbered choices of environments shaped by cultural preferences, then the issue may be better explained as a problem in political economy rather than one of cultural relativism. Unfortunately, we did not appreciate this dilemma sufficiently in constructing our survey to provide a means by which to disentangle this possibility.

The third variable in the sampling scheme—stages in the family cycle—also did not appear to be as strong a factor in explaining differences in experience or preferences. In most cases where we controlled for the effect of population groups through an analysis of variance, we found the difference attributable to stages in the family cycle to be quite small and not statistically significant. Where a significant difference did exist, it was usually between the elderly and the rest, and it usually concerned aspects of the environment clearly related to retired life, age-related incapacities, and the like.

In summary, according to our findings, difference in income class remains the single most important variable in explaining the quality of residential experience and judgments about what constitutes a good place to live. This must be the primary consideration in developing a city- or regionwide strategy of managing and improving the quality of residential envrionments. The issues related to family-cycle differences and to ethnic differences, if any, may be significant only when considering a specific site with a specific group of users.

SUMMARY AND CONCLUSIONS

What can we conclude from a review of our findings from these three approaches? First, we believe that, by selecting different angles of inquiry, we were able to obtain from our respondents a far richer picture of their residential experience and perceptions than we would have obtained from any single approach. Clearly, these three approaches tapped different dimensions of residential perception, not all of which were incorporated within the neighborhood unit concept.

Second, we found a number of perceptions at variance with or supplementary to the neighborhood unit concept. These suggest that we need a variety of approaches to residential design to encompass the variety of perceptions. The neighborhood net, if cast again, should be cast wider to include regional concerns

Third, despite the differences in perceptions, and despite the failure of the neighborhood unit concept to tap salient attributes of residential life, the social meaning of neighborhood is important, even if the transformation of that meaning into three-dimensional form has continued to elude designers. This conclusion reinforces our second conclusion. The issue is an important one in people's minds, and to dismiss neighborhood values as being erroneous or as being founded on false beliefs about people's behavior in residential areas (as numerous critics have done) is to deny the validity or merit of alternative approaches and design paradigms that would compensate for some of the mistakes and omissions in the neighborhood unit concept.

We might also ask at this stage what, if any, implications for policy can be drawn from the way in which responses about the residential milieu, form, and setting varied by income, ethnicity, and stages in the family cycle. Throughout the preceding three chapters, and in discussions earlier in this chapter, we have consistently emphasized income as the most dominant factor in explaining major differences in residential experiences and perceptions. In an overall sense, our impression of the relationship among our three independent variables and the three different categories or responses can be summarized as shown in Table 6.1.

The differences attributed to income, therefore, reflect economic differences between population groups as being the most pronounced, rather than ethnic/cultural differences or stage in the family cycle. As was alluded to earlier, this finding both simplifies the kinds of design solutions that might be proffered and makes the kernel of the problem more resistant to solution. If age and stage in the family cycle were the major problems, they nevertheless are ones that all population groups will face; hence, the solutions would apply to each groups. If race/ethnicity/cultural differences were the most pronounced, then clues to alternative design solutions could be derived from the expressed wants of the respondents. In fact, the comparatively recent recognition on the part of planners

TABLE 6.1. An Interpretive and Global Summary of Findings vis-à-vis the Three Main Independent Variables or Sample Characteristics

	Response variability by sample characteristics		
Levels of residential experience	Income	Ethnicity	Stages in family cycle
Residential milieu	Strong	Moderate	Weak
Residential form	Strong	Weak	Weak
Residential setting	Strong	Weak	Moderate

and developers of population and age diversity has already resulted in residential development differentiation (e.g., retirement communities, housing developments and programs for the elderly; condominiums for childless parents or singles), although, admittedly, a middle or upper income usually is necessary to capitalize on these special developments.

But the income differences noted in our findings suggest solutions that aim at redistributing the resources in our society more than they suggest strictly design alternatives. We have already mentioned the importance of design in determining cost. This point is emphasized when we recognize that residential development is the result of the joint production of the public and the private sectors. A subdivision laid out in large lots, with wide streets and substantial land dedicated to parks, schools, libraries, and community centers, will carry with it expensive land costs. Because developers operate on rough rules of thumb regarding the ratio of development to the cost of the land, we can be sure that the houses and community facilities provided by the developer will also be expensive, assuring a very expensive final product. Only a small segment of our society will be able to afford that joint design decision between the public and private sectors. Other subdivisions and housing developments will be built to lower standards, but it is the cost of the developments (both new and "used" through "filtering") that will largely determine the settlement pattern in urban regions, although discrimination (both overt and covert) will be influential as well. But these decisions based on cost will be the most important arbiter of the types of people who live in these developments—the prime source of residential satisfaction or dissatisfaction. Furthermore, the income patterns that develop in these residential areas will, in turn, influence shopping-center patterns, resulting in a greater variety and usually a greater proximity of stores (except for the very rich) for the middle- and upper-income groups. And variety and proximity are additional sources of residential satisfaction.

Thus, we see that the distribution of resources in our society is more influential in residential satisfaction than design standards. Design alternatives can help ameliorate the pernicious effects of completely unregulated development, but they are limited in ensuring more than a minimal level of the residential satisfaction that can be attributed to design.

As we see it, the differences in the quality of the residential experience at all three levels, arising from income differences, lead to the questions of equity and "basic needs." That is, the relevant prepolicy question is whether these current levels of residential experience are equitable. Are they a result of some systematic inequity in the way that public resources are allocated to residential services and environments? If so, what levels of compensatory allocation of public resources (Lineberry, 1975) will be needed, for whom, and in what order? If not, is it still possible to establish the "basic requirements" of a good residential environment as the target for future residential planning, irrespective of income?

The differences in the residential experience (and therefore in preferences) arising from ethnic and life-cycle differences, however weak, raise a different type of policy issue. Here, it seems, diversity and choice are more critical dimensions of public policy than are equity or basic needs, although for the elderly, who are critically dependent on their immediate setting, the notion of "basic requirements" may still apply. Finally, the goal of diversity and choice may lead to another thorny issue about specific standards. That is, should standards be the same or different for different user groups? Can public policy set anything but uniform standards? Are exceptions equitable? Although these questions have usually been the province of political scientists and economists interested in public service delivery (Rich, 1979), we will respond to them within the general framework of residential planning and design in Chapter 7.

NOTES

1. Although there is some empirical evidence that, all else being equal in terms of social milieu, microspatial factors can influence social relations, neighboring, and the like (see, for example, Dyckman, 1961; Festinger, Schachter, and Back, 1950; Kuper, 1951).
2. Critics, however, are not totally convinced of Newman's findings and proposals (see Mawby, 1981; Merry, 1981).
3. Nevertheless, the return of the "ma and pa" type of store in the form of the Seven-Eleven, the Stop and Go, and other chain convenience stores suggests that they serve a function that is not yet outmoded, despite the revolution in shopping-center developments.

7

Toward a New Design Paradigm

At the outset of our study, we suspected that the neighborhood unit model might have outlived its purpose, despite its persistence as a paradigm. This skepticism was reinforced by strong and forceful arguments made by various scholars and practitioners, which we reviewed in Chapter 2. Indeed, we have found little in our own research to deny this skepticism, much less to refute the critics of the model. True, some of our findings lend credence to certain facility location standards linked to the neighborhood unit concept, but that credence is not sufficient to warrant a wholesale endorsement of the model. Nor for that matter does the fact that the ideal of the neighborhood is still highly valued by the majority of the respondents mean total vindication of the neighborhood unit concept. All it suggests is that some of the contextual values still have a following, even though the manifest functions of residential environments have become diversified, as shown by our findings and those of others in recent years.

So, is it now time to bid farewell to the neighborhood unit? Or should we be more circumspect in the face of enduring "neighborhood" ideals and simply say, *"The neighborhood unit is dead, long live the neighborhood"?* This approach would allow us to endorse the neighborhood values and ideas but would let the model that pursued these values lapse while alternative paradigms are embraced. We are convinced that the design profession needs the kind of guidance provided by the neighborhood unit concept. However, the profession needs other guides from which to choose as well. What follows is a case for such an alternative paradigm of residential planning and design.

THE SEARCH FOR AN ALTERNATIVE

As the most commonly accepted concept of the residential environment—the neighborhood—has proved inadequate as the sole basis for a physical planning and design concept, a fresh start is in order. In our inquiry into the nature of residential experience, we have searched for some consistent patterns and some unifying themes that could tell us as simply as possible about the meaning of the residential environment in people's lives. We had hoped in our study to discover some global, all-encompassing sense of the residential area, in the same way that the meaning of a house was interpreted by Martin Buber (1969) as a person's defense against "the uncanniness of the universe" and "the chaos that threatens to invade him" (quoted in Alexander *et al.*, 1977, p. 393); or as described by Gaston Bachelard (1969) as a "nest," a "sanctuary"; or, as more recently suggested by Lee Rainwater (1966), as a "haven"; or by Clare Cooper (1971) as a "symbol-of-self."

Can any of these concepts apply to the definition of the residential area as well? Perhaps. We have seen, in Chapter 3, that some of these meanings did appear to underlie open-ended descriptions of the residential area. Indeed, this is to be expected; after all, the residential area is the next larger realm of life experience (Campbell *et al.*, 1976), which subsumes the house. Still, we do not feel that the concepts that seem to have so eloquently captured the meaning of the house encompass the overall sense of the residential environment in its everyday usage and purposes.

Indeed, as we survey our findings, our immediate impression is that no single meaning applies to the residential area within a population group (much less between groups), despite the differences that we have reported. Some see it as being territorially defined, others as being a social milieu. To some, it is a "haven"; to others, it is nothing more than a vague transition between one's home and the larger community—even the city. The spatial extent of the residential area varies dramatically—from a street block or an apartment complex to a whole section of the city. Some feel that the residential area and the neighborhood are the same; others see these two entities as being quite different. Some define a residential area as functionally homogeneous and exclusive; others define it as a functionally heterogeneous and all-encompassing area. Even people's sense of what environmental settings and hardware should belong to the residential area is quite different. We conclude that the *residential area means different things to different people,* even within the same population group.[1]

Our conclusion, so baldly stated, may seem anticlimactic. Is this commonplace conclusion the essence of the last 4 chapters? We think not. We have merely oversimplified here to make a point: the rigidity, inflexibility, insensitivity, and unresponsiveness of the neighborhood unit concept impose decided limitations on its ability to satisfy a diversity of needs. Our findings do not suggest that it has failed, only that it is frequently insufficient. Certainly, we need a fresh approach to grappling with its current inadequacies from the abundant hint that we find of its increasing irrelevance in the final two decades of this century, when major shifts in lifestyles are emerging. In formulating a new approach, we look first to some previous efforts for guidance and then address ourselves to some fundamental shortcomings in these approaches before describing our own alternative.

THE PREVIOUS SEARCH FOR ALTERNATIVES: A REVIEW

The design profession has not been totally oblivious of the criticisms of the neighborhood unit concept. Over the years, a few designers and planners have sought to develop frameworks for city design that did not rely on the neighborhood unit as the main "building block" (Herbert, 1963). Some of these ideas were developed—and remain—as mere suggestions; they have had relatively little impact on actual city development. Others have been implemented and tested at least once. All have attempted to break away from the bondage of the neighborhood unit concept.

David Crane (1960), for example, argued that the public infrastructure is the more permanent element of city form, and that therefore it, rather than residential cells, should be used as the main organizing framework for a city designed for change and adaptability (see Figure 7.1). A similar theme of permanence versus

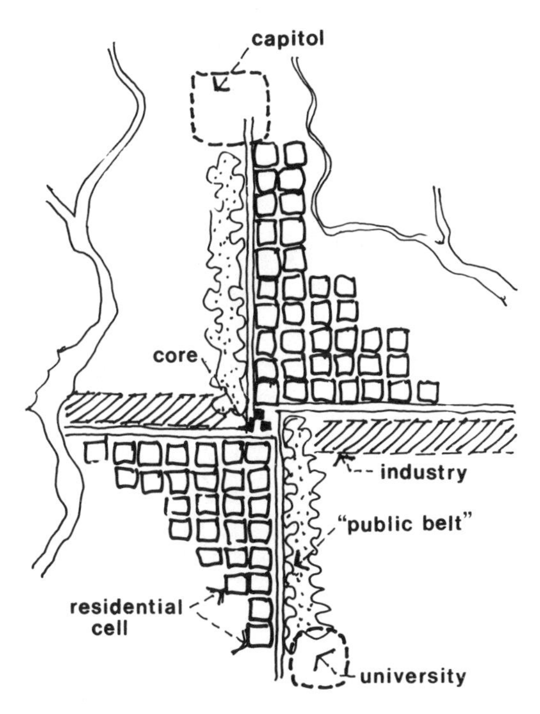

FIGURE 7.1. A hypothetical redesign of Chandigarh, India, illustrating Crane's argument. Adapted from D. Crane "Chandigarh Reconsidered". *AIA Journal* 5 (1960): 33–39, by permission of David Crane and *AIA Journal.*

change was the basis of a framework for city design (and redesign) proposed by Allison and Peter Smithson (1967; see Figure 7.2). Both of these concepts rejected the neighborhood unit as being too large, too specific, and too inflexible. Instead, smaller residential clusters were emphasized on the ground that they have a greater "social validity" (Herbert, 1963). It should be noted, however, that both of these schemes were responding to the perceived need of the early 1960s to design a city that could accommodate rapid physical change and obsolescence. The design thinking in that decade was preoccupied with the efficiency and the adaptability of the city form rather than with the questions of livability and equity that became the main concerns of the 1970s. Another scheme that did focus on the livability question as the basis for city design was the plan for the British New Town of Hook (see Figure 7.3). Although Hook was never built, its design philosophy influenced the subsequent generation of British

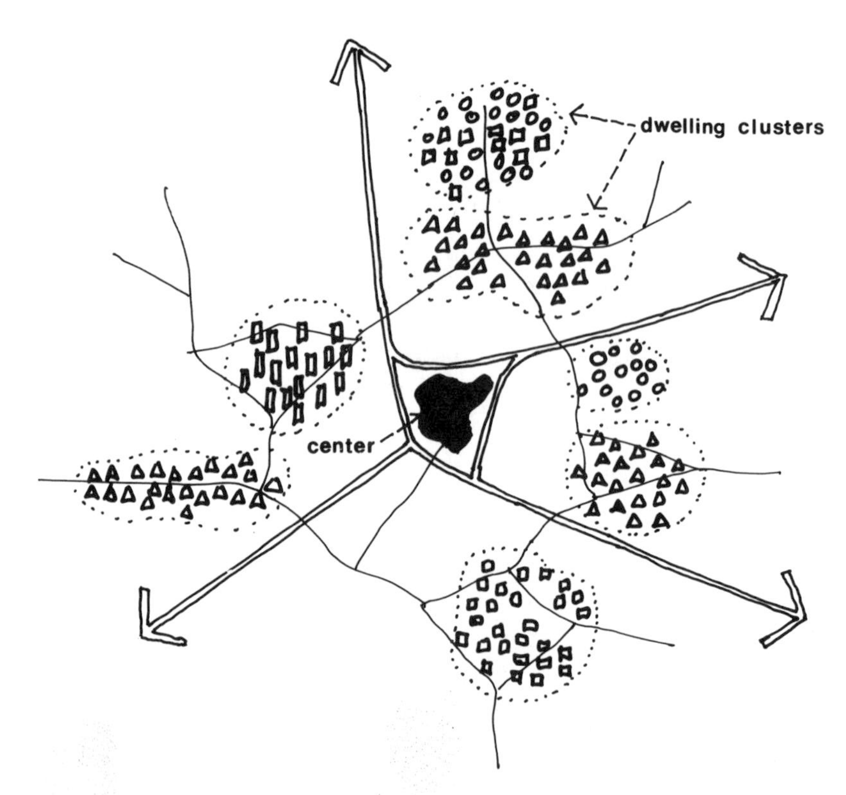

FIGURE 7.2. "Cluster City"—a hypothetical city design without the neighborhood unit concept. Adapted from A. Smithson and P. Smithson, *Urban Structuring.* Copyright 1967 by Van Nostrand Reinhold Co. Used with the permission of Van Nostrand Reinhold Co.

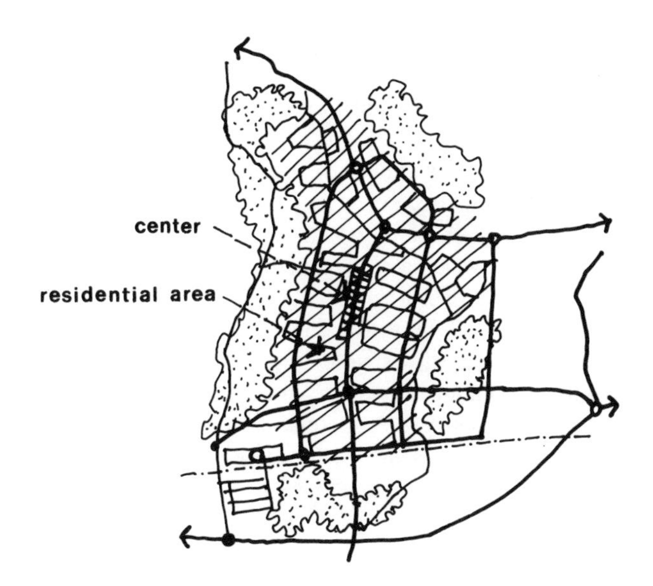

FIGURE 7.3. The plan of Hook (unbuilt)—another case of a city designed without the neighborhood concept. Adapted from London County Council, 1961.

New Towns, especially Cumbernauld. The Hook plan suggested a major departure from the neighborhood unit concept by focusing on small "cul-de-sac" housing groups, because they were "small enough for a group to be recognisable both as a social and as a visual unit" (London County Council, 1961, as quoted in Herbert, 1963). Choice and diversity in housing composition were also stressed in the plan. Boundaries between residential groupings were de-emphasized, and their interconnectedness was highlighted by the sensative location of public services and facilities within easy access of the residential area. These principles were essentially implemented in the design and development of Cumbernauld, which was to become perhaps the first British New Town to escape the influence of the neighborhood unit concept (see Figure 7.4).

The neighborhood unit, once a hallmark of the British New Towns, was further avoided in subsequent New Town development in Britain. In the case of Milton Keynes, for example, the physical form of the city was developed on a curvilinear grid of arterials intersecting at 5/8- to 2/3-mile intervals (see Figure 7.5). The resulting cells—roughly 250 to 300 acres each—are called *environmental areas* and contain a variable mix of housing, open spaces, shopping, schools, and other "activity centers." The planners stressed that these "environmental areas" are not to be construed as "community" or "neighborhood" because "there is little or no evidence available that would suggest that a particular size of unit or grouping of units is more or less appropriate socially or administratively'' (Milton Keynes Development Corporation, 1970, II, p. 305). This scheme, the planners emphasized, is predicated on maximizing ''freedom of social development, movement and choice.'' Thus, Milton Keynes has come to symbolize a final and definitive disavowal of the neighborhood unit concept in contemporary British New Town development. Its cohorts, such as Runcorn and Washington, have followed suit.

The idea of the neighborhood unit has also been rejected by socialist planners in favor of a two-tier schema for organizing a residential area. Basing their ideas largely on considerations of service delivery rather than community, one group of Russian planners (Gutnov, Babunor, Djumenton, Kharitonva, Lezava, and Sadovskij 1968) suggested a smaller residential complex of 1,500–2,000 persons as the basis for locating the primary-level services that should be as close to home as possible, and a larger residential sector of 25,000–35,000 persons that would include larger, nonprimary service centers appropriate at this scale

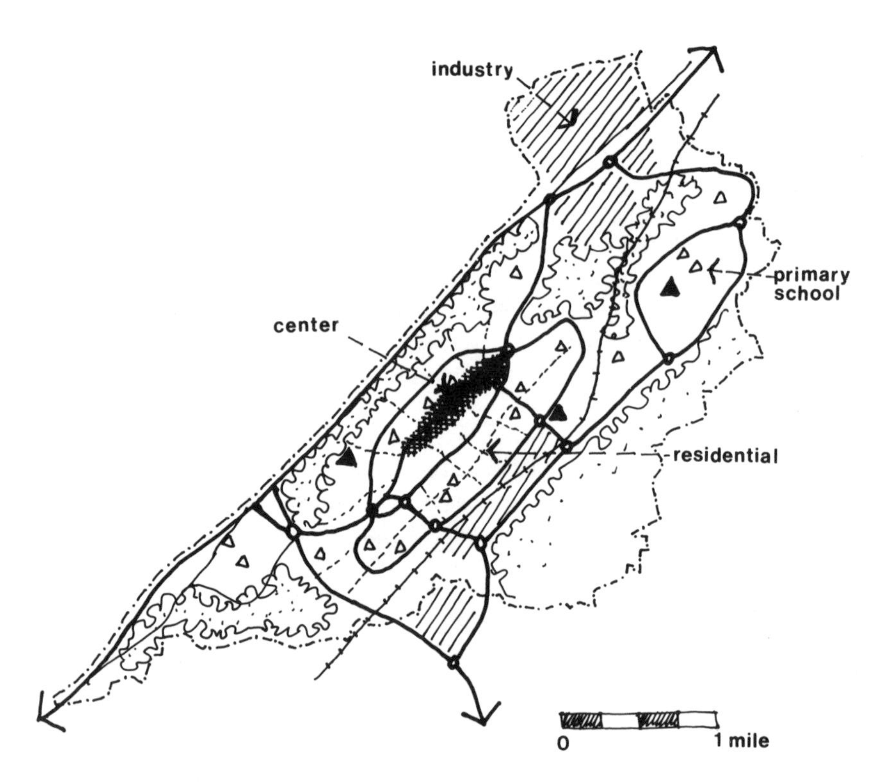

FIGURE 7.4. A schematic sketch of Cumbernauld—inspired by Hook and without any explicit consideration of neighborhood units. Adapted from Cumbernauld Development Corporation, 1968.

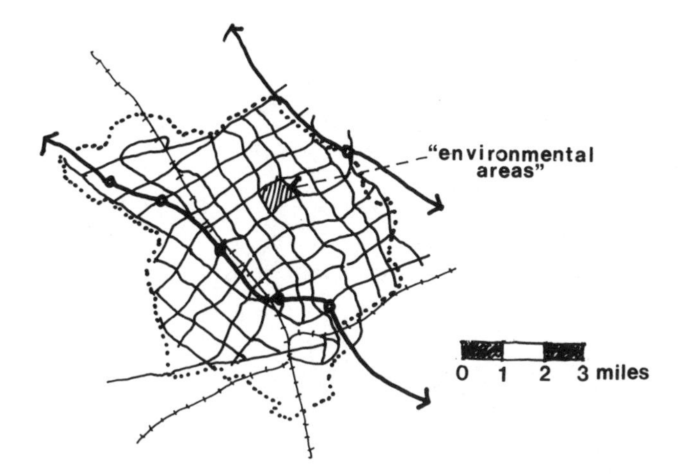

FIGURE 7.5. A schematic sketch of the Milton Keynes concept—again, a disavowal of the neighborhood unit. Adapted from Milton Keynes Corporation, 1970.

(see Figure 7.6). Although the neighborhood unit was formally rejected here, it was, in fact, merely supplanted by a hierarchy of cellular modules not unlike the modification of the neighborhood unit concept proposed previously by Gropius (1945), Forshaw and Abercromble (1943), and Gibberd (1953).

An innovative, albeit theoretical, schema for residential planning has been proposed by Hendricks and MacNair (1970), in which the neighborhood unit is replaced by a concept of residential place designed to accommodate variations in lifestyle, personality, and stages in the life cycle (see Figure 7.7).[2] These authors conceptualized this residential place as a mosaic of smaller, homogeneous cells designed to service such variations in the user profile. This scheme was expected to maximize choice and diversity at a coarser grain, yet to preserve the social or ''lifestyle'' homogeneity at the cell level. The authors saw their model of residential places as being responsive to the social problems and pathology that arise from role discontinuity and place incongruence as people change or move through different stages of the life cycle. Like the ones proposed by Crane and the Smithsons, this schema is also concerned with the adaptive qualities of the environment and, at the same time, emphasizes aspects of psychosocial well-being.

Finally, Nicholas Low (1975) advocated what he called a ''non-centrist urban structure'' in which the provision of residential services is based on maximizing consumer choice and social equity. He argued that traditionally the normative models of urban form have been conceptualized as nested hierachies of service centers in which the lowest common denominator is a residential area

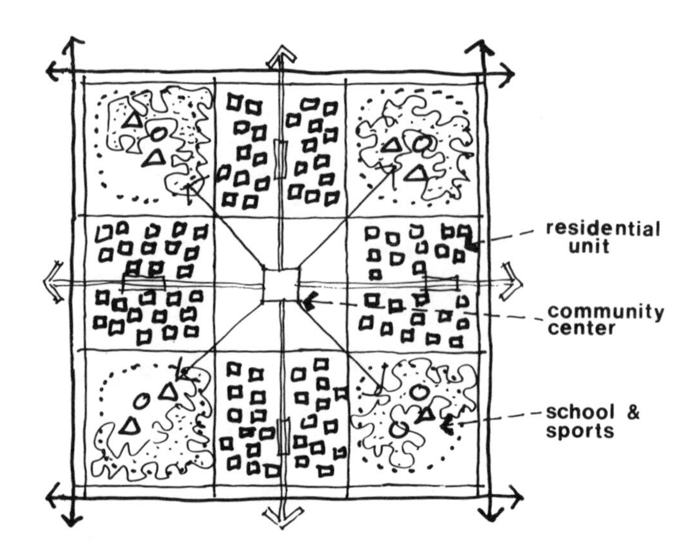

FIGURE 7.6. A socialist alternative to the neighborhood unit concept. Adapted from Gutnov *et al.*, 1968. From *Ideal Communist City* by George Braziller. Copyright 1971 by George Braziller, Inc. Adapted with permission.

with a service nucleus at its core (Alexander, 1964), not unlike the neighborhood unit. He maintained that this "centrist" approach to city design is based on such conventional wisdom as a central-place-type theory of the market, consumer perception of convenience, commercial efficiency, community ideals and cultural values, and the like. Centrism in city design, he argued, leads to an inequitable

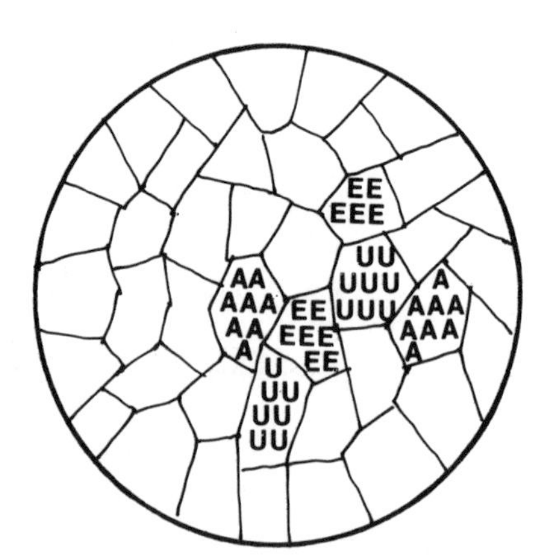

FIGURE 7.7. A mosaic of socially homogeneous subcommunities The random letters indicate different social groups. Adapted from F. Hendricks and M. McNair, 1969.

distribution of services; fails to reflect the true operations of the market; is structurally rigid; and is inimical to consumer sovereignty. As an alternative, Low pointed to the South Hampshire Study by Colin Buchanan and Partners (1966) as a design for noncentrist urban structure (see Figure 7.8). This scheme is characterized by a transport network of a rectangular road grid; a continous service reservation adjacent to this grid; a supply of redundant land; and community control over land-use decisions at the local level. This approach, Low said, is expected to maximize both consumer sovereignty and equity in service provision.

This review of alternative models of residential planning would be incomplete if we did not consider a recent scheme of community design proposed by sociologist Gerald Suttles (1975). Suttles's perspective is somewhat different from the ones discussed above, for he has focused on the social and political implications of community and jurisdiction, rather than on the physical place *per se* (see Figure 7.9). Nevertheless, he was concerned about such issues as class and racial segregation, equity in the distribution of public services, and patterns of community participation in metropolitan areas—all of which are relevant to the planning and management of future residential environments.

In his proposal, Suttles suggested a three-tier ordering of the residential community in metropolitan areas. At the lowest level is the "minimal" subcommunity, which is either the smallest named residential unit or a less defined residual area that exists between discrete, identifiable residential units. The next

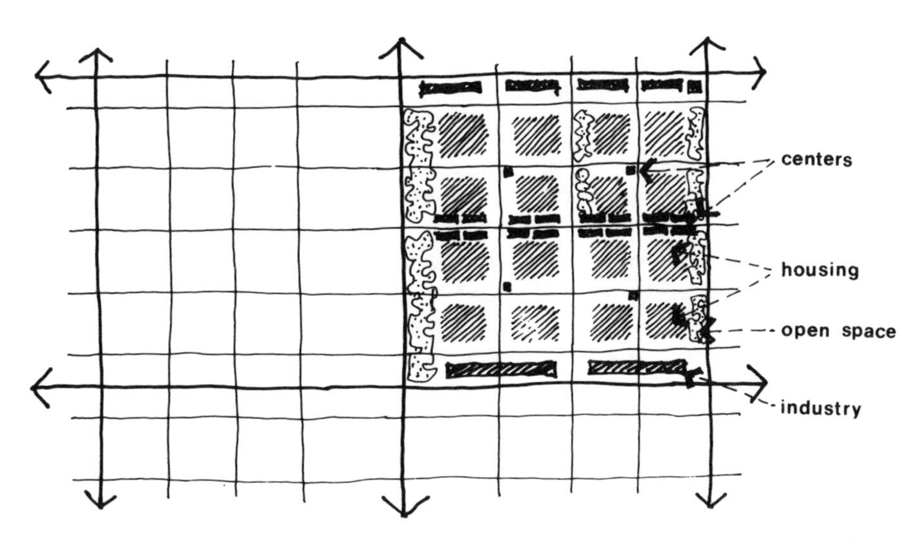

FIGURE 7.8. The "noncentrist" design—plan of South Hampshire, England. Adapted from C. Buchanan and Partners, 1966.

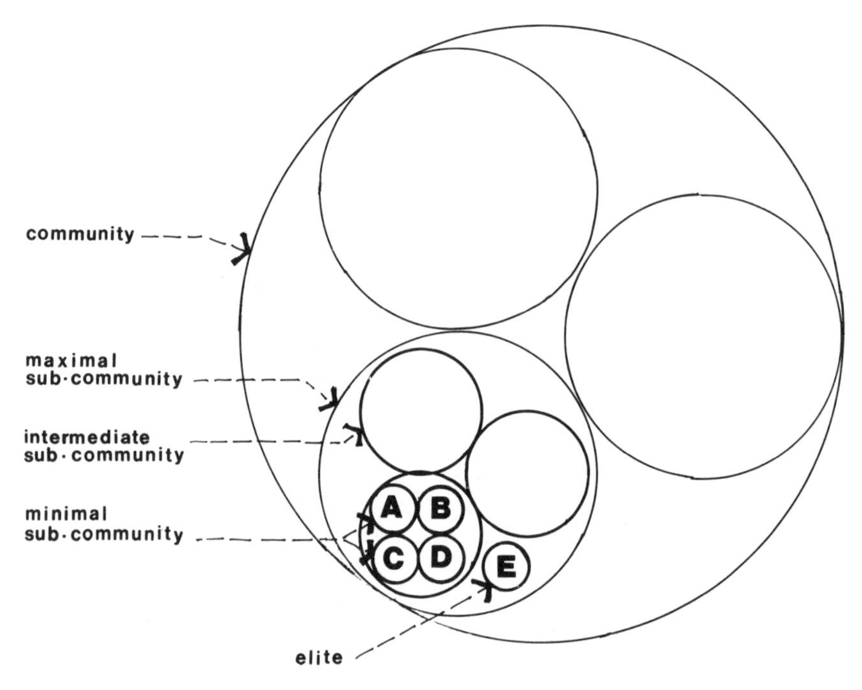

FIGURE 7.9. Suttle's schema for community design. The letters A, B, C, D, and E indicate different minimal sub-communities.

level is a larger subcommunity composed of these minimal communities but emphasizing a heterogeneity that is lacking at the small level. At this level, communication, interaction, and community participation are possible between diverse racial, socioeconomic, and ethnic groups. Finally, Suttles proposed the highest level, or the "maximal" subcommunity, large enough to support several public-service agencies and to allow for choice and differentiation of services for clients. This level of community would include not only a number of intermediate levels but also at least one "elite" community, so that total integration of all social strata is possible to this level of decision making. Suttles argued that the subcommunities must be given clear identities at the lowest level, and that the district boundaries of the public agencies should be redefined to match the subcommunity boundaries. This schema, Suttles claimed, would lead to an efficient distribution and consumption of public services through some decentralization, increased choice and competition, and some voluntarism. Although one might argue that rigid boundaries and compartmentalization of urban areas may interrupt the evolving social ecology of the city and thereby freeze current inequities of the urban social order (Harvey, 1973), Suttles's schema provides an important institutional framework for managing the quality of the residential environment.

CONSIDERATIONS FOR ALTERNATIVE FORMULATIONS

These previous efforts all fall short of our expectations, and a case for alternative formulations can still be made. Our study (and others from the social sciences) suggests that some innovation is necessary and possible, but that any new formulation must be guided by considerations of what we know today about the nature of the collective residential life experience.

1. The inherent variability and complexity of residential life experiences elude easy conceptualization in terms of hierarchy, organization, and order. Yet, as we have seen earlier, most of the previous design concepts convey a strong belief in hierarchy—in the idea that order can be imposed on complexity only through levels or classes. This belief system is, of course, not backed by any research findings; it reflects only the originator's presumption that such a scheme will prove beneficial. But we are not very comfortable with such expert judgments these days. One reason is that it is an elitist, a "from-the-top-down," approach, that presumes superior knowledge on the part of the expert; such an approach can lead from the view that people *will* behave in a physical environment in the way that the designer intended to the view that people *must* behave in the way that was intended. Ultimately, it can become a subtly autocratic view, unpalatable to a democratic society. But more important, there is reason to believe that the simplification that necessarily comes with a hierarchical organization is the result not of some special insight or inspired design but often of the cognitive limits of human minds in coping with a complex information structure. Alexander (1965) has argued quite effectively that, whereas the "natural city" is a complex mesh of overlapping and interconnected sets of human relationships, the "artificial cities" (the new towns, for example) that we design are typically conceptualized as mutually exclusive or nested sets of hierarchical relationships. Whereas the real city represents a "semi-lattice" type of organization, our representations of it usually take the form of a "tree"-like schema. Alexander argued that designers are all hopelessly "trapped in a tree" and "cannot achieve the complexity of the semi-lattice in a single mental act" (p. 60).[3]

If, on the other hand, we presume that in complex situations people are the best determinants of order, and that the very complexities present do not yet warrant deference to expertise, then an approach that allows people to devise their own residential-area configuration might make up for the shortcomings and insufficiencies of current design paradigms. In this view, future design frameworks for residential environments may have to be constructed with reference to many different opportunities, activities, desire lines, "behavior circuits," consumption choices, and the like. In this schema, no one particular school, supermarket, church, gas station, park, or clusters thereof may become the organizing basis for a residential environment; rather, it is the overall range of choice, the alternative settings, activities, and places within the larger context, that become important. That is, a person may define his or her residential area not just as including the nearest supermarket or drugstore, but also as including other similar drugstores and supermarkets that might be visited for comparative shopping. In this view, the larger urban context acts as a ''cafeteria'' of choices in activities, facilities, and opportunities, rather than as the community kitchen ladling out a single fare. Personal tastes, lifestyles, and means form an individual's activity pattern, which, in turn, dictates a stable circuit of behavior (Perin, 1970). It is this selection of facilities and activities from the range of available choices that defines the residential ''action space'' (Horton and Reynolds, 1971; Wolpert, 1965). Literally, millions of possible behavior circuits and alternative action spaces exist for the residents of a particular neighborhood. Although many of these behavior circuits and desire lines overlap and intersect, most of them remain distinctively separate and independent of each other. As a result, it is almost impossible to obtain a consensus on any territorially defined ''neighborhood'' or residential area.

2. *The sense of neighborhood or community may not necessarily be attainable through the physical design of the residential environment.* The neighborhood-unit-concept focus was largely physical, with some presumed social outcomes; efficiency was a secondary consideration. As our review of the historical basis for the oganization of residential areas reveals, however, religious, military, and administrative foci are also possible, as well as equity, which is a comparatively recent focus. But the immediate focus of our concern here is whether a social construct can be forced on the residents by manipulating physical means. Social scientists have repeatedly warned us that it is not possible. There is very little in our findings to suggest otherwise. By separating our inquiry into three levels of experience, we have shown that the constructs of the social milieu are not particularly dependent on the constructs of the residential form. We have seen that, for the majority of our respondents, the ''neighborhood'' and the ''residential area'' were not the same entity, and that the transition from one to the other in people's minds was vague and fuzzy at best. In light of all this, we would like to suggest that it is necessary to make a clear break between the goal of establishing a sense of community and the goal of creating a good place. Perhaps, we should concede that the goal of directly establishing the sense of community or neighborhood is simply unattainable through physical design, and indeed, it need not be a goal of physical design anymore. For we still know too little about the role of the physical space, if any, in shaping social space. Furthermore, place and community may be two different domains of life experience, and although the two may coincide on occasion, they are not necessarily contingent on each other.

This is not to say, however, that we cannot provide a physical setting with the desired environmental hardware and activity settings from which individuals can derive their own sense of community, or that we cannot focus on the performance characteristics of the residential milieu, such as safety, air quality, amenities and convenience, place qualities, and services, without worrying about whether a sense of community can be achieved.

3. A rigid, cellular, building-block approach may be antithetical to the more open ended, unstructured, and evolving nature of the residential life experience. The examples that we have reviewed, as well as the neighborhood concept, all subscribe to the same assumptions: a cellular construct is desirable; a building block that in combination with other building blocks can be assembled into a larger community, city, or urban region is ideal. There is more than a suggestion of modular thinking here, with its attendant uniformity, monotony, and ultimate rigidity, despite claims of flexibility. Worse, a modular orientation subtly reinforces the need for some higher authority to decide on the construct, for once chosen, all shapes and forms (and people) must conform or be left out of the design.

This building-block approach also presumes that urban existence can be organized only by resort to discrete forms. This approach stresses separation, boundaries, apartness, and even mutual exclusivity. But our research and that of others show overlapping, interwoven lines of relationships among people, places, and activities that transcend and defy such artificially constructed boundaries. An alternative approach would take account of continua, as well as the lack of a clear demarcation between the activities in our urban lives.

Finally, the building-block approach results in designs that tend toward stasis rather than flow, and toward an immutable design rather than a concept that accommodates the inevitable flux and change of urban areas. We need to concentrate on an approach that is less wedded to a particular form and more concerned with facilitating individually desired residential behavior, regardless of the configuration of the track or trace that that behavior circuit leaves in its wake.

4. A territorially well-defined unit focally organized around an exclusive set of commercial and community facilities may not be the best, or the only, residential form. In the most frequently adopted cell-like organization, the cells themselves are seen as having a model internal structure of their own (a predictable extension of the organic analogy), as epitomized in the neighborhood unit formulation. The focal organization is expected to create a sociopetal space, to induce social interaction and communion, and to facilitate a shared community identity. But for the center (i.e., the shopping and community facilities) to function, a minimum population or areal size is necessary, either one of which is usually much larger than the current average increment of new or recycled development in a metropolitan area. But more important, this particular internal structure of the cell has to assume that the activity patterns of the residents must flow toward the core. Although this idea may have had some validity in the days of Clarence Perry, in today's urban area it is an untenable assumption. To the extent that cognitive maps reflect people's "behavior circuits" (Perin, 1970), our findings suggest that people's activities and interactions are much too open-ended (and perhaps much too spontaneous) to be dependent on such a contrived physical arrangement. Furthermore, even for service delivery, such arrangements are not very efficient, as some critics have pointed out (see Chapter 2). We also feel that such arrangements respond to such fundamental consumer-behavior traits as comparative shopping or variety-seeking activities. We have argued elsewhere that access to alternative opportunities for shopping, leisure, recreation, or socialization are valued in today's society more than mere proximity to one set of services and facilities. Indeed, this preference is rather effectively reflected in the decisions about the location of present-day shopping centers, chain stores, small retail outlets, and even such quasi-public facilities as churches, temples, and clinics. In such an environment, this particular model of residential cells has very little appeal or validity. What is important is how these cells are linked to the existing (and evolving) streams or conduits of private and public services and opportunities in an urban area.

5. *The new formulation must take into consideration equity in the distribution of residential resources for different groups.* The uniformity inherent in the designs that we have reviewed, at the first glance, promulgates equality. But we have seen how equal design can have unequal effects with respect to, say, the elderly, and how any given design has a cost that can be afforded better by some groups than by others. Are equal standards for all really equitable when incomes are not? We believe that the design paradigm must provide for unequal design (inputs) that will result in more equal outcomes for different groups. Thus, design provisions or service levels may have to be increased substantially beyond the minimum standards to compensate for the deficiencies now entrenched in many residential areas. In education, compensatory programs are offered to those pupils who are handicapped developmentally or socioeconomically, The same could be done to obtain equity in the distribution of residential quality—by providing compensatory services and facilities for environmentally "handicapped" residential areas.

We recognize, of course, that this is easier said than done. In a way, in our new formulations, we are looking for a basis for allocating "impure public goods" (Harvey, 1973)—public services and the like (indeed, according to David Harvey, even a supermarket is an impure public good). And if the demand for these impure public goods is regarded as being inelastic across income groups—which we are willing to assume—any purportedly egalitarian formulation would amount to a redistribution of real income. Indeed, this is what the application of unequal standards biased in favor of the impoverished—or, for that matter, even the application of equal standards in the face of existing inequality—would mean. Although the redistribution of income is accepted in certain sectors of general welfare (health and education being two main examples) it still remains politically unpalatable in any form (although, as Harvey, 1973, argued, the existing allocation of resources through the market mechanism has already generated an implicit income redistribution in favor of the affluent).

But if we are to speculate on the spatial format of such a redistributive formulation of residential planning—begging, for the time being, the question of political feasibility—what should it be like? We know that if all impure public goods were treated as private goods available through a pricing mechanism established in a free and perfectly competitive market, the spatial form of the location of residential services would probably have approached a hierarchy implicit in the Löschean model of spatial equilibrium (again, see Harvey, 1973, pp. 87–91). And indeed, it is apparent, in retrospect, that the neighborhood unit concept as a service delivery module was trying to imitate this concept of efficiency, albeit in a haphazard manner. But as we argued in the previous section, not only did the neighborhood unit fail to achieve any efficiency, it also inadvertently contributed to spatial inequity by creating some exclusive domains of service usage. On the other hand, if such services and facilities were to be located instead in the interstices between residential cells, so that they were not within the exclusive domain of any cell, a greater range of choice and availability—and possibly some degree of equity—could be reached (Low, 1975).

6. The new formulations must accommodate the changing context of urban development and the increasing variety of residential environments. When the neighborhood concept was first formulated, and subsequently when it was being legitimized as quasi-official doctrine by local agencies and adopted as the basis for *Planning the Neighborhood* (PTN), the orientation toward growth was largely directed to the suburbs. Large subdivisions like Levittown were the rage, and there was considerable talk of new towns or other forms of large planned developments. Much of this development has come to pass, and the neighborhood concept was incorporated into these developments.

Today, the prospects for development are more mixed and varied. The development of new towns as such has ceased because of their high initial capital costs. Although large planned subdivisions will still be built, their growth will be slowed by stricter subdivision and environmental regulations and by reduced demand because of energy and transportation costs. Financing costs have further contributed to a decline in the pace of development. On the other hand, inner-city renewal, both public and private, has accelerated considerably. Gentrification has introduced a more piecemeal, less comprehensively planned redevelopment, and the current stress on infill development suggests the variety of contexts in which new development will take place. The rehabilitation of existing areas is still another form of development that does not conform to the requirement of large parcels of raw land so necessary for implementing the neighborhood concept in its pure form. Thus, any new formulation must fit the variety of development contexts that will be arising in the future.

Finally, we must consider the likely prospect for higher densities of residential units. The high cost of housing and the increased cost of accessible locations suggest that many households will be willing to accept increased density in exchange for lower costs. Townhouses, garden apartments, high-rise residences, and the many forms that condominiums take are all more prevalent today than when the neighborhood concept was first formulated. Moreover, the growing use of planned unit development means that the physical designer has greater freedom than was allowed by traditional subdivision and zoning regulations.

All of these considerations suggest the need for increased flexibility in design concepts and design standards to satisfy the preferences of the population. Although greater flexibility promises ultimately to fit residential needs better, it also introduces considerably more complexity, as we will see.

7. Any new formulation must also consider the relationship between different areas and the larger urban system. In the neighborhood unit formulation, the transition from the neighborhood scale to the larger scale—despite an assumed hierarchial order—was insufficiently formulated. Discontinuities in scale, function, and hierarchy were never properly resolved. The major emphasis was on the cells; the interstices were ignored. Proponents of the neighborhood unit concept had very little to say about the leftover spaces between the neatly defined neighborhood units. Such an oversight should be avoided in developing new concepts.

The neighborhood unit was not particularly focused on the issues of the larger community, or at least on those attributes of the local environment that are influenced by the larger urban or metropolitan development patterns (such as air quality). But urban life is increasingly taking on metropolitan dimensions, which are now evident at the neighborhood level, and new design constructs must recognize that changing reality.

In considering the larger urban framework, new formulations must also be mindful of the dynamics of neighborhood change. The framework should be adaptable to such aspects of the ''filtering'' process as different income or ethnic groups' moving through an area, the wholesale internal migration of ethnic groups, or internal mutation within a residential area, such as rental or condominium conversions, ''doubling up,'' and the like.

In summary, our proposed formulation takes its authority from research findings but does not attempt to impose a social order on people's lives by manipulating the physical environment. It seeks to break with the tradition of formally designing a sense of social community by providing an environment that will serve as a facilitative setting in which the inhabitants choose their own configuration. The formulation places increased emphasis on flow, movement along continua, and the acknowledgment of interwoven, overlapping spaces that render discrete but linked adjacencies less important. The organization of this formulation should be horizontal, periphery-seeking, and "sociofugal," with places for polynodal centers of attraction. Its composition and configuration should be in a form flexible enough to meet the particular needs of user groups, even if this flexibility means different compositions for different groups, and it should also be polymorphous enough to fit into existing urban areas as well as to become the basis for new ones. Finally, it should lend itself to serving metropolitanwide needs as well as local residential-area ones.

A PROPOSAL TO MEET THESE CRITERIA

What would a proposal organized along the lines that we have suggested look like? This is a natural question, but we must be cautious in providing an answer because there is danger in responding at all. The original neighborhood unit concept was proposed as a means of achieving social ends. But in the process of translating social, political, and moral concerns into a design for three-dimensional space, the concept became increasingly emphasized as an end in itself. To be sure, there is always a problem in determining where in the means–ends chain one wishes to focus attention: one person's means is another person's ends—and still a third person's means to a different end. We believe that satisfaction with residential life is largely (but not entirely) achieved through treating the design of residential space as a means for facilitating that satisfaction, but that implementing that design is not synonymous with the achievement of that satisfaction.

Thus, we believe that a detailed proposal merely falls into the same trap as the neighborhood unit concept. Its very specificity may drive out alternatives from consideration. In a more basic sense, a detailed, graphic representation of these criteria unwittingly misses the significance of our urgings. It immediately becomes a static representation of a dynamic circumstance; it tends to demarcate and delineate what are inherently vague lines. We must be careful, therefore, to provide only the rudiments of a prototype, allowing the designer and the context of the moment to flesh out the more specific details necessary for that situation.

Nevertheless, with these caveats in mind, we can sketch some of the alternative configurations that such an approach would have. These are shown in the accompanying sketches (Figure 7.10). Their main features are as follows: (1) the schemes suggest an optimal nexus between dwelling units and the public and private facilities deemed necessary in a residential area: (2) they are not based on a hierarchy or building-block type of organization; (3) the corridors or nodes of services and facilities, rather than the residential areas or dwellings clusters, are seen as the basis for the spatial framework of residential planning; (4) however, these services and facilities do not belong to the exclusive domain of any ter-

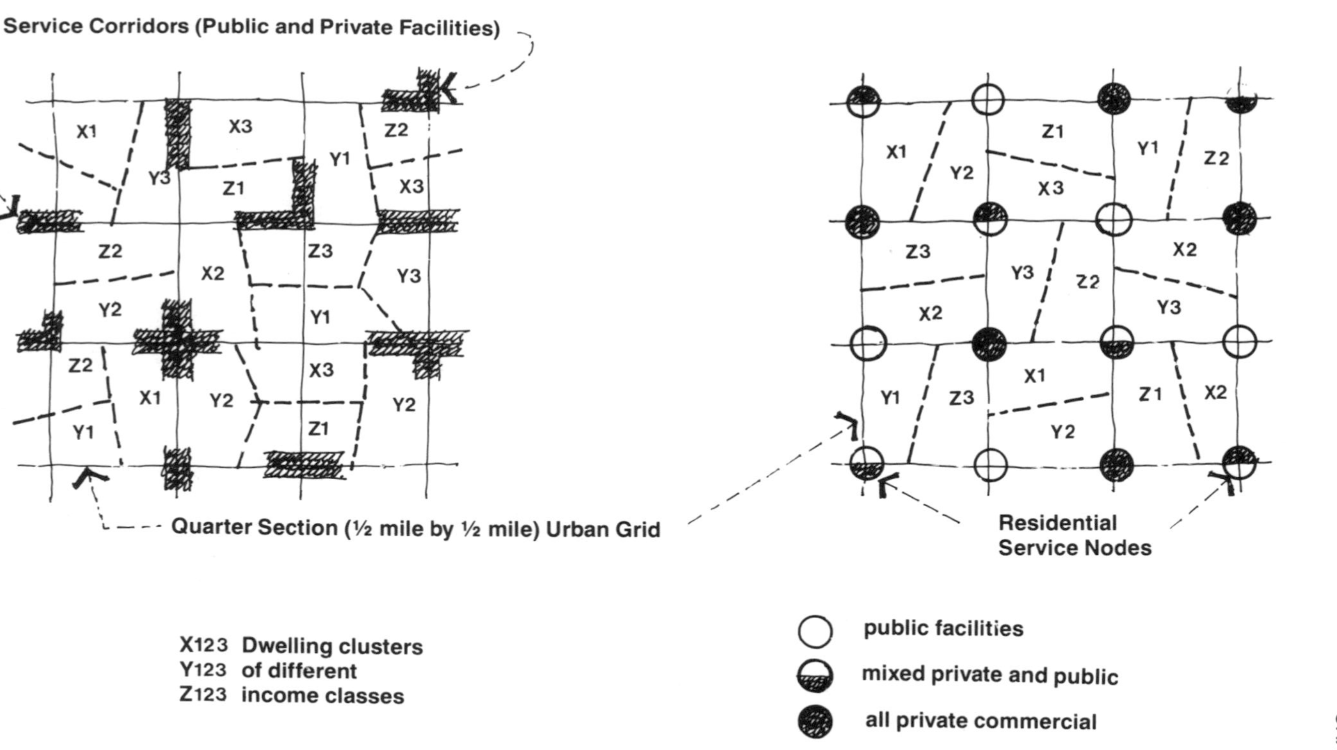

FIGURE 7.10. Two schematic alternatives illustrating the proposal discussed here.

ritorial residential unit; (5) the exact mix and capacity of the services and facilities included in the nodes or corridors are meant to reflect the density and mix of the residents in proximate locations; (6) although discrete, territorially defined, and homogeneous residential "cells" are not intended, the possibility of a mosaic of "dwelling clusters" within the residential areas is acknowledged; (7) these dwelling clusters, however, can be seen essentially as a collection of dwelling units, which, by design, density, affordability, or some other commonly shared site or physical characteristics, form a homogeneous physical unit; (8) although these clusters may come to represent some social homogeneity arising from similar incomes or stages in the life cycles of their residents, no presumption is made about their social cohesion or "neighborhood"-ness.

We include the last three features having in mind the social ecology of existing cities and some practical considerations of residential planning and design. Thus, the purpose of these features is simply to: (1) acknowledge that residential areas are not physically, and therefore sociologically, monolithic and that different dwelling clusters are likely to be inhabited by different social groups; (2) address the variable service needs and preferences of different social groups residing in different dwelling clusters; (3) recognize that these are likely units of future transformation and change (McKie, 1974) and are therefore targets for different strategies for neighborhood improvement (Downs, 1981; Goetze, 1976); and (4) suggest that these physical units represent the most likely and feasible increments of the spatial mobility of different social groups through such well-studied processes as "filtering," "gentrification," "displacement," "doubling up," renewal, or subdivision of land. The location, distribution, and mix of these dwelling clusters are important considerations in a framework of residential planning and design.

Although the above features of our spatial utopias would seemingly satisfy most of the requirements that we discussed earlier in this chapter, some nagging questions still remain. First, how do these configurations serve equity? Or alternatively, do these schemes have to be modified somehow to meet the equity criterion? Second, how do we get from here to there? In other words, what are the essential processes through which such desired states can be approached? In some ways the answers to these two questions are related, and we will address them both at the same time.

Leaving these spatial utopias aside for the time being, an equitable distribution of residential environmental services and facilities can be achieved, at least in the abstract, through different, but not necessarily mutually exclusive, approaches. These can be called (1) the democratization of urban form; (2) the remedial allocation of public resources; (3) the ghettoization according to the "public choice" paradigm; and (4) the deghettoization of social groups. These views are not original with us (although they may not be known commonly by the labels we use here), and each view has a constituency of scholars, academics, and professional planners. Let us consider the implications of these four approaches for the illustrative spatial frameworks that we proposed previously.

The Democratization of Urban Form

This approach is inspired by the spirit of social, political, and environmental equality. It envisages an urban form that exists for the enjoyment and benefit of all. Lynch (1981) can be considered a proponent of this philosophy; he emphasized access and control as two important characteristics of a good (efficient and equitable) city form. An accessible city form not only provides easy access to all of its resources and facilities but also assumes equal mobility for all. Control implies not only political equality but also, in this case, one's right to shape one's immediate environment.

We also include Sennett (1970b) as an advocate of this approach, although from a somewhat different position and with a different prescription for urban form. He has argued that the planners' quest for spatial order has resulted in the isolation of social groups, and that, in order to promote democratic discourse between diverse groups, a certain amount of "congestion" and "disorder" is necessary in the mixture of uses, facilities, and services. Sennett's prescription for urban form also suggests compact living and an "overload" (or planned surplus) of public and private facilities.[4] Although ostensibly designed for increasing the intensity and diversity of social encounters, this prescription can be interpreted to mean equalization of access to city resources for all groups.

Access, control, disorder, congestion, overload of facilities, and the like can thus be seen as some of the basic tenets of democratic urban form, although it may not be possible to lump them all together. Conflicts invariably will arise. Disorderly cities are not very accessible, for example, and vice versa. There are some other conceptual and practical difficulties with this prescription. An accessible city form by itself may not result in an equitable distribution of opportunities, in the face of unequal mobility, which is a result basically of income inequality. Decentralization of planning and management decisions resulting from local control of residential environments may lead to some disorder, redundancy, or congestion. But a desired "overload" of facilities, to the extent that this means planned surplus of settings and facilities, is not likely to happen either through normal market mechanisms or through public resource allocation in a time of fiscal austerity. Although the link between disorder, congestion, and oversupply, and environmental equity (if not equality) may remain largely a matter of speculation, it can be argued that transferring the control over design decisions to the local level—"a bottom-up approach"—may affect the outcome positively, because it will be based on local priorities rather than on some universal standards.

In any event, how do these criteria for democratizing urban form fit with, or alter, our earlier utopias? On the one hand, our schema is general enough so that most of these criteria can still be applied. On the other hand, the local control of design decisions may lead to some conflicting developments. For example, it is possible that locally based planning decisions may lead to a new "morality of isolationism" (Sennett, 1970a), and thus to a regressing to cellular identities and exclusive control of public facilities. After all, these are what characterizes small, local communities in a metropolitan area. This will be clearly in conflict with some of the goals of our illustrative utopias.

The Remedial Allocation of Public Resources

In reviewing the relevant literature, Rich (1979) concluded that empirical research into the intrajurisdictional distribution of public services—such as education, library services, police protection, recreational facilities, and street maintenance—shows both "patterned and unpatterned inequalities, in the objectively measured distribution of services," although not necessarily biased against the poor.[5] But as he pointed out, these studies looked at input measures (what he called "policy output") such as the neighborhood design standards, rather than output measures (or what he called "policy outcomes") or the residential satisfaction of neighborhood residents. He argued that instead it is the inequality in policy outcome that should be the focus of research on equality. Indeed, equal standards may not mean equal outcome and certainly may not advance the cause of equity. This view essentially supports our earlier argument that unequal design or standards may be necessary to ensure equal outcomes.

The inequality that we have documented in our research is essentially one of output or "policy outcomes" as seen through the eyes of the residents. To equalize these differences[6]—that is, to bring the environmental qualities of the poor areas up to par, even by their own standards and expectations[7]—would require the remedial allocation of public resources. The application of equal standards in the name of "equal protection" clearly will not do. Poor areas where there is much fear of crime and violence may require a substantially higher level of police protection than is needed in other areas. Although this higher level means unequal services, it would result in a more *equitable* distribution because the recipients would be "in a more nearly equal life circumstance after receiving the services than before" (Rich, 1979, p. 152).

This standard of equity is based on the concept of "equal results" (Levy, Meltsner, and Wildavsky, 1974, pp. 16–18) and approaches the ethical system championed by Rawls (1971).[8] But if the remedial allocation of public resources for improving environmental facilities or services in the poor areas is possible only at the expense of a reduction of services in other areas (and this might be particularly true for services other than capital improvements), the other equity standard,[9] "pareto optimality" (which is blind to the history of a particular social formation), will no doubt be violated. And this violation will be considered unacceptable according to the libertarian traditions of the Western democracies.

A general limitation of the remedial approach, as in the case of the previous one, is that it makes no explicit provision for the services and facilities that are normally supplied by the private sector. Even if it is possible, say, to furnish the poor neighborhoods with neighborhood parks, bicycle paths, street lights, and the like, to substantially augment fire and police protection, and to eliminate undesirable land uses and environmental features by public decree, the list of setting deprivations (see Chapter 5) will still remain quite long. Can public policy force a supermarket or a drugstore chain to open a facility in the heart of a poor neighborhood?

Finally, can our spatial utopias be reached through this remedial approach? We think not, even though its larger objectives may not be in conflict with our proposal. The remedial allocation of public resources, by itself, is not likely to be enough to advance the cause of equity in the distribution of residential quality resources. It will simply create refurbished ghettos of the poor, as accomplished by the Model Cities program of the Great Society days.

Ghettoization According to the Public Choice Paradigm

For the lack of a better term, we call this third approach to equity a *ghettoization strategy,* although public choice theorists would no doubt object to such a characterization. The concept of equity, according to this view, is largely seen as one of "institutional fairness," where equity in service delivery is not an end in itself but a means to the end of an equitable distribution of life chances (Rich, 1979). Here, considerable emphasis is given to institutional responsibilities and equal access to institutions that are capable of satisfying service demands. The public choice theorists would argue (Bish, 1973, 1975) that service equity—by means of the "institutional fairness" doctrine—can be achieved only when "political units are organized so that the beneficiaries of a service are simultaneously the group which determines the nature of that service and pays for it" (Rich, 1979, pp. 387–388). At the metropolitan level, this suggests that the smallness and the homogeneity of the local community may best provide a "good fit" between service demand and supply and to eliminate the problems of the "limited divisibility" of public goods and the "free riders" phenomonon. It is also claimed that under these circumstances, the prospects of voluntarism and "coproduction" benefits—where consumers indirectly contribute to the efficient production of a service, such as by refraining from littering or by mobilizing a voluntary cleanup campaign (Rich, 1979)—can be vastly increased.

Intrajurisdictionally, this approach amounts to neighborhood control of the production and the consumption of local services and thus smacks of a fourth level of government. Indeed, Rich (1977) offered an elaborate schema for institutional design that shows how such political units can be organized and how they would interact with each other. In this model of the decentralization of service production, the monopolistic nature of public service production is replaced by a more competitive, free-market type of environment. The public choice theorists recognize, of course, that the poor neighborhoods will not have enough revenue or resources to support the production of their own services and that subvention from some other higher source of government will be necessary. So redistribution of income will still be in effect, but public services will not be the mechanisms of such a redistribution. In the final analysis, their approach is really a question of efficiency in the classic economic sense, that is, how to make the public sector behave as if it were a market economy. The argument of equity by "institutional fairness" can be seen as a subterfuge.

But what are the social implications of such a development? Neiman (1975) correctly argued that autonomy based on homogeneity, as advocated by the public choice theory, is often protected by land-use control and zoning practices, as is common in exclusive suburban and no-growth communities (Frieden, 1979). Similar practices can very well be adopted at the neighborhood level to keep "undesirable" people out and thus limit the physical mobility and the civil liberties of many. They could permanently freeze the residential differentiation and thus create perpetual ghettos. Cities can indeed get polarized into a "public city" of service-dependent populations (Dear, 1979; Wolch, 1979), on the one hand, and an affluent "private city" of conspicuous consumption, on the other. It can breed the isolation, alienation, and mutual fear that Sennett (1970a, b) bemoaned. Indeed, this scenario is the polar extreme of the democratic urban form that both he and Lynch (1981) advocated.

Because this model focuses only on the distribution of public services, it offers few answers to the question of how equity can be achieved in access to other facilities and services produced in the private sector. Indeed, it remains mute even on the impact of the ghettoization model of service delivery on the location and the movement of private sector establishments. Like the previous two, this approach, by itself, fails to offer a route map to the utopias that we presented initially.

The Deghettoization of Social Groups

The fundamental problem that must be recognized in approaching questions of equity is that the ecology of urban opportunities and the ecology of the urban poor are not spatially coterminous. Although these noncoterminous ecological formations are explained by the classical economic theories as inexorable outcomes of the bid–rent function, in reality these are well protected by many nonmarket institutional and political barriers. Exclusionary zoning, covenants, subdivision regulations, urban renewal, and, indeed, the neighborhood unit formula (recall the arguments of Bauer, 1945; Isaacs, 1948a, b, c; and Wehrly, 1948; in Chapter 2)—all of our familiar planning tools—have been a party to these protective efforts.

Thus, the scenario of polarized cities, which we rejected in our earlier discussion, already exists in many instances. Scholars have described it as the phenomenon of *dualism:* two independent circuits of economy, market, and social processes. In the context of residential planning and design, the problem can be defined in terms of *spatial dualism* of public services and consumer opportunities. And as long as this dualism exists, the problem of equity in spatial design can never be addressed adequately through the remedial approach, the institutional design of service delivery systems, or the democratization of urban form.

So, what is needed is a strategy that can mitigate this spatial dualism as much as possible in order to maximize equity in the distribution of urban services and consumer opportunities. And given the present constraints of the market and the public delivery system, one way to achieve this goal is to move the deprived population near the existing nodes and corridors of such facilities. This approach amounts to "breaking up" the ghettos—probably at both ends of the income spectrum, but particularly at the poor end. This deghettoization strategy does not necessarily mean a fine-grained integration by class, ethnicity, or stages in the family cycle, which is neither possible nor desirable, as many social scientists have argued strictly on social or behavioral grounds (Booth and Camp, 1974; Gans, 1961; Greenbie, 1974). Instead, the scenario of deghettoization that we would like to project is one of a coarse-grained mix but of a grain substantially finer than what we see today. As we indicated earlier, individual dwelling clusters may provide the optimum scale for mixing of this kind.

Deghettoization is not a new idea. Others have advocated it for reasons of equality, equity, and opportunities for human development. Anthony Downs (1973) and the Davidoffs and Gold (1970, 1971) have argued that the deghettoization process can begin only when the protective barriers of the affluent suburban communities are removed. Some progress in this direction has been made after many years of legal battle. Recent trends of inclusive zoning—to increase the supply of low- and moderate-income housing through the private sector—have already started a certain amount of finer grained mix by income class. Rent control measures have, in some cases, stopped or at least slowed down class segregation in existing heterogeneous communities. Gentrification, on the other hand, has had a deghettoizing influence in some instances as middle- and upper-income groups have moved back to the inner-city neighborhoods. If the displacement problem associated with gentrification can be avoided, or matched by replacement housing for the same class of people in other neighborhoods, the revitalization programs may be made less undesirable for the poor. (See Sumka, 1979, and Hartman, 1979.)

Finally, we must be mindful of some dramatic changes that are under way in the established patterns of housing consumption. The disappearance of reasonable fixed-mortgage rates in home financing and the increasing costs of housing are making home ownership increasingly difficult for newly formed households. If high interest rates continue, the investment appeal of real properties will no doubt drop. The incentive for home ownership may be further eroded if the no-deduction, flat-rate income tax proposal being discussed by the present administration comes into being. In any event, as more and more American households are forced to join the ranks of renters, or to buy only much smaller housing spaces (such as condominium apartments, two- or three-family shared houses, and mobile homes), the traditional opposition to compact and mixed residential areas may disappear. Changes in the economy, in the role of women, and in family structure may force other necessities to the forefront and make mixed-grain residential areas more functional, as in many Third World cities. Thus, working mothers may no longer find suburban environments an ideal setting and may prefer to live in areas where baby-sitting services and domestic help are nearby. Indeed, a deghettoization scenario may not be very utopian after all.

A COMBINATION OF STRATEGIES

Although the last approach offers the best fit with our spatial utopias, an optimal developmental scenario would include elements of all four of the approaches just discussed. A citywide (or metropolitan-area-wide) residential planning strategy will begin with a remedial allocation of public resources (as, seemingly, this is a central bureaucratic decision) to the worst of impoverished and deprived areas. These improvements in public facilities and services are then likely to bring these areas to a minimum threshold of acceptability. Some of these areas might then be ready for some revitalization—gentrification without displacement—and an influx of residents from higher levels. While this phase is under way, a citywide or regionwide strategy of deghettoization of the poor can be undertaken. An obvious possibility is using inclusionary requirements as an "overlay zoning" on existing land-use patterns, and enforcing these requirements on private or public subdivision or redevelopment efforts. As these processes begin, accessibility to city resources will increase for all, and so will individuals' choices of and control over their activity circuits. With a mixture of consumer groups, commercial establishments will no doubt diversify, thus increasing the opportunities for social interaction between social groups. Thus, we see that democratization of the urban form is possible once the deghettoization process is complete. Finally, once a stable state is achieved, efficiency in the consumption and the production of public services can be maximized by local organization and control, based, in this case, not on homogeneity but on some other naturally evolving coalition of groups sharing an area of common interest.

CONCLUSIONS

We can summarize this alternative approach by recasting it in the framework that we used in Chapter 2 to explain the genesis of the neighborhood unit concept.

With respect to the question of values, we are assuming that the new *contextual* values will be largely individual. Rather than creating a device to enforce elite values or to impose them on society so as to mold better citizens or at least to save them from the ravages of the unplanned and unregulated city, the newer values of physical and mental well-being take a more individualistic approach. Rather than designing residential areas in some particular fashion to accomplish some larger social ends, the new approach is: How do we design the the residential area so as to achieve well-being in each resident? This is more of a "bottom-up" view of the problem (and its required solution), in which the physical and mental well-being of each individual is seen as leading, by aggregation, to the well-being of society. The other approach assumed that if we provided an environment that was good for social ends, then we could be assured that the individuals in that society would be well provided for as well. The very fact that only one design was recommended suggests, perhaps, an unconscious desire for social uniformity, whereas the more recent search for alternative lifestyles and concomitant design suggests greater attention to individualism and permissiveness.

This new individualistic approach, in turn, suggests our own *manifest* formulation. If we emphasize networks, communication channels, and pathways for flow and movement, we emphasize the neighborhood in flux rather than in repose. Given this framework, the residents are better able to organize their own residential areas in the ways that please them. This ability may allow them to develop a sense of a "roving neighborhood" (Reimer, 1950), rather than a stationary one. Rather than society's dictating the form from above, by means of prescribed and uniform standards, individuals can construct and reconstruct their personal form from below. We provide the skeleton and the nervous system; they flesh it out to suit their needs and purposes. Furthermore, we provide a framework that does not impede their own individual construction (in the way that too definite boundaries or a too rigid a formulation might). But we let them put it together in ways that are meaningful to them.

We have proposed that, within this individualistic framework, the *tacit* values of residential organization must emphasize equity. Individual opportunities for life chances and a decent residential environment cannot be improved for all without some equitable distribution of the environmental settings, the amenities, the facilities, and the services that are normally desired in residential areas. Although our manifest formulation envisages equity in distribution by showing an optimal distribution of different dwelling clusters with respect to networks, corridors, and nodes, this scenario can be reached only through some deliberate changes in the existing urban form. We have focused on some possible strategies for achieving equity and have presented our recommendations.

What we have proposed may not be a final prescription. There may be other ways of responding to these contextual, manifest, and tacit values. But this is relatively less important than the values themselves. If these values are acceptable, institutions and designs will no doubt come to accommodate them.

NOTES

1. The fact that we did not elicit any one construct, or even a dominant construct of residential area from our respondents, suggests at least three possibilities. First, this outcome simply may be an artifact of our methodology (e.g., our research assumptions, sampling plan, or interview protocol), and not a function of the phenomenon itself. But other studies using more narrowly defined research variables (cf. Lee, 1968), and based on different methodologies and nationally drawn samples (Coleman, 1978; U.S. Department of Housing and Urban Development, n.d.), have also found a wide range of perceptions and constructs of the residential environment, many of which are similar to ours.

   A second possibility is that our findings were influenced by the environment in which our study was conducted (i.e., Los Angeles). The variability in responses may simply reflect the wide variations in the physical character of residential locations in the Los Angeles metropolitan area. An alternative explanation in this vein would suggest that the formless, characterless sea of undifferentiated sprawl in Los Angeles precludes the formation of a dominant construct, or that the high degree of intracity residential mobility—which is typical of Los Angeles residents—contributes to these "multiple visions" of residential environments. All of these explanations seem plausible, but we suggest that Los Angeles is typical of most metropolitan development in the West and the Southwest and of much of the newer suburban development in the South and the East, and so this "bias" is certainly not irrelevant.

   We think the most likely explanation is that personal constructs of the residential environment are likely to be highly variable because of differences in individual values, tastes, lifestyles, personality traits, and the like, and that this is why we have seen so many different versions of what the residential area means to people and what matters most to them in this environment.
2. Alexander *et al.* (1977) have further developed this concept, proposing specific criteria for physical design.
3. We hasten to add, in all fairness to the design profession, that the tendency to impose order and organization in the face of complexity is not unique to designers; others are known to succumb to this human frailty. It will be recalled that sociologist Suttles's (1975) schema for a community design framework also showed a neat hierarchical organization.
4. See, for example, "Richard Sennett Lectures on Democratic Theory and Urban Form" (1982).

5. But empirical research into citizens' perceptions of differences in neighborhood services shows rough equality (the respondents were asked to rate their neighborhood services with those of other neighborhoods), although many researchers doubt the validity of such perceptions as a surrogate for obejctive measures (Rich, 1979). There are other reasons to question these findings: neighborhood pride, variable expectation levels, variable thresholds of satisfaction (Campbell *et al.*, 1976), limited knowledge about the technical aspects of public-service delivery systems, and the like can all distort such perceptions. In any event, these findings need not be seen as contradictory to ours, because we did not query our respondents about specific neighborhood services or other comparative qualities. We were more interested in the global perceptions and evaluations of the residents' current environments.
6. Public choice scholars would, of course, argue that equality of outcomes is extremely difficult and costly to measure, and that therefore such a goal is not particularly viable (see Rich, 1979).
7. That the poor have lower levels of expectation—and therefore of satisfaction—has been shown by Campbell *et al.* (1976). It has been evident in other parts of our study—especially in relation to attribute satisfaction and trade-off preferences. The implication is that bringing the environmental qualities of poor areas up to the residents' expectations may still fall far short of any objective equality outcome, as measured by the incidence of crime or by the level of air pollutants.
8. Schulze (1980) interpreted Rawls's argument, that social welfare is improved when the welfare of the worst off improves, as essentially an expression of egalitarian ethics.
9. We are omitting the notion of "market equity," which, even according to the public choice theorists, would amount to an exacerbation of existing inequalities (Rich, 1979).

References

Alexander, C. *Notes on the Synthesis of Form*. Cambridge: Harvard University Press, 1964.

Alexander, C. "A City Is not a Tree." *Architectural Forum* 4, (1965): 58–62; 5 (1965); 58–61.

Alexander, C., Ishikawa, S., Silverstein, M., Jacobson, M., Fiksdahl-King, I., and Angel, S. *A Pattern Language: Towns, Buildings, Construction*. New York: Oxford University Press, 1977.

American Institute of Architects. *The First Report of the National Policy Task Force*. Washington, D.C.: Author, 1972.

American Public Health Association, Committee on Hygiene of Housing. *Planning the Neighborhood*. Chicago: Public Administration Service, 1960 (rev. ed.).

Appleyard, D. "City Designers and the Pluralistic City." In L. Rodwin and Associates, *Planning Urban Growth and Regional Development*. Cambridge: M.I.T. Press, 1969.

Appleyard, D. *Planning a Pluralist City: Conflicting Realities in Ciudad Guyana*. Cambridge: M.I.T. Press, 1976.

Bachelard, G. *The Poetics of Space*. Boston: Beacon Press, 1969.

Baer, W. C., and Banerjee, T. "Behavioral Research in Environmental Design: Beyond the Applicability Gap." In *The Behavioral Basis of Design*. Book 2, *Selected Papers*, edited by P. Suedfeld and J. Russell. Stroudsburg, Pa.: Dowden, Hutchinson and Ross, 1977.

Banerjee, T., and Flachsbart, P. G. "Factors Influencing Perceptions of Residential Density." In *Urban Housing and Transportation*, edited by V. Kouskoulas and R. Lytle, Jr. Detroit: Wayne State University, 1975.

Banerjee, T., Baer, W. C., and Robinson, I. M. "Trade-Off Approach for Eliciting Environmental Preferences." In (Ed.), *Man-Environment Interactions: Evaluations and Applications*, Edited by D. Carson. Stroudsburg, Pa.: Dowden, Hutchinson and Ross, 1974.

Barker, R. G. *Ecological Psychology*. Palo Alto, Calif.: Stanford University Press, 1968.

Bauer, C. "Good Neighborhoods." *The Annals of the American Academy of Political and Social Science* 242 (1945): 104–115.

Berger, B. M. "Suburbs, Subcultures and the Urban Future." In *Planning for a Nation of Cities*, edited by S. B. Warner, Jr. Cambridge: M.I.T. Press, 1966.

Bish, R. L. "Commentary." *Journal of the American Institute of Planners* 39 (1973): 403, 407–412.

Bish, R. L. "Commentary." *Journal of the American Institute of Planners* 41 (1975): 67, 74–82.

Booth, A., and Camp, H. "Housing Relocation and Family Social Integration Patterns." *Journal of the American Institute of Planners* 40 (1974): 124–128.

Buber, M. *A Believing Humanism: Gleanings*. New York: Simon and Schuster, 1969.

Buchanan, C., and Partners. *The South Hamshire Study*. London: H.M.S.O., 1966.

Bunker, R. "Travel in Stevenage." *Town Planning Review* 38 (1967): 213–232.

Burby, R. J. III, and Weiss, S. F. *New Communities, USA*. Lexington, Ma.: Lexington Books, 1976.

Buttimer, A. "Social Space and the Planning of Residential Areas." *Environment and Behavior* 4 (1972): 279–318.

Campbell, A., Converse, P. E., and Rodgers, W. L. *The Quality of American Life: Perceptions, Evaluations and Satisfactions*. New York: Russell Sage Foundation, 1976.

Carr, S., and Schissler, D. "The City as a Trip: Perceptual Selection and Memory in the View from the Road." *Environment and Behavior* 1 (1969): 7–35.

Chapin, F. S. *Human Activity Patterns in the City: Things People Do in Time and in Space*. New York: Wiley, 1974.

Churchill, H. S. "How to Prevent Neighborhood Decay." *Journal of the American Bankers' Association* 7 (1945): 52–53.

Coleman, R. "Attitudes toward Neighborhoods: How Americans Choose to Live." *Working Paper No. 49, Joint Center for Urban Studies of the Massachusetts Institute of Technology and Harvard University*, 1978.

Committee of Regional Plan of New York and Its Environs. *Regional Survey of New York and Its Environs. Vol VII: Neighborhoods and Community Planning*. New York: 1929.

Cooper, C. *The House as a Symbol of Self*. Berkeley: Institute of Urban and Regional Development, University of California, 1971.

Crane, D. "Chandigarh Reconsidered." *AIA Journal* 5 (1960): 33–39.

Cumbernauld Development Corporation. *Cumbernauld Preliminary Planning Proposals*. Cumbernauld, Scotland, 1958.

Dahir, J. *The Neighborhood Unit Plan—Its Spread and Acceptance*. New York: Russell Sage Foundation, 1947.

Davidoff, P., Davidoff, L., and Gold, N. M. "Suburban Action: Advocate Planning for an Open Society." *Journal of the American Institute of Planners* 32 (1970): 66–76.

Davidoff, L., Davidoff, P., and Gold, N. M. "The Suburbs Have to Open Their Gates." *The New York Times Magazine* 11 (1971): 39–42.

Dear, M. "The Public City." In *Residential Mobility and Public Policy*, edited by W. A. V. Clark and E. G. Moore. Beverly Hills, Calif.: Sage, 1979.

deChiara, J. D., and Koppelman, L. *Planning Design Criteria*. New York: Van Nostrand Reinhold, 1969.

deChiara, J. D., and Koppelman, L. *Urban Planning and Design Criteria*. New York: Van Nostrand Reinhold, 1975.

Dewey, R. "The Neighborhood, Urban Ecology, and City Planners." In *Cities and Society*, edited by P. K. Hatt and A. J. Reiss, Jr. , New York: Free Press of Glencoe, 1961.

Downs, A. *Opening Up the Suburbs: An Urban Strategy for America*. New Haven, Conn.: Yale University Press, 1973.

Downs, A. *Neighborhoods and Urban Development*. Washington, D.C.: Brookings Institution, 1981.

Dyckman, J. W. "Of Men and Mice and Moles: Notes on Physical Planning, Environment and Community."*Journal of the American Institute of Planners* 27 (1961): 102–104.

Everitt, J., and Cadwallader, M. "The Home Area Concept in Urban Analysis: The Use of Cognitive Mapping and Computer Procedures as Methodological Tools." In *Environmental Design: Research and Practice, One*, edited by W. Mitchell. Los Angeles: University of California, 1972.

Federal Housing Administration. Land Planning Bulletin No. 1. *Successful Subdivisions: Planned as Neighborhoods for Profitable Investment and Appeal to Home Owners*. Washington, D.C.: Superintendent of Documents, 1941.

Federal Housing Administration. *Underwriting Manual*. Washington, D.C.: Author, 1947.

Festinger, L., Schachter, S., and Back, K. *Social Pressures in Informal Groups*. New York: Harper, 1950.

Fischer, C. S., Jackson, R. M., Stueve, C. A., Garson, K., Jones, L. M., and Baldassard, M. *Networks and Places: Social Relations in Urban Setting*. New York: Free Press, 1977.

Flachsbart, P. G., and Phillips, S. "An Index and Model of Human Response to Air Quality." *Journal of the Air Pollution Control Association* 30 (1980): 759–768.

Forshaw, J. H., and Abercrombie, P. *County of London Plan*. London: Macmillan, 1943.

Fried, M. "Grieving for a Lost Home." In *The Urban Condition*, edited by L. J. Duhl. New York: Simon and Schuster, 1963.

Fried, M., and Gleicher, P. "Some Sources of Residential Satisfaction in an Urban Slum." *Journal of the American Institute of Planners* 27 (1961): 305–315.

Frieden, B. J. *The Environmental Protection Hustle*. Cambridge: M.I.T. Press, 1979.

Gallion, A. B., and Eisner, S. *The Urban Pattern: City Planning and Design*. New York: Von Nostrand, 1975.

Gans, H. J. "Planning and Social Life: Friendship and Neighbor Relations in Suburban Communities." *Journal of the American Institute of Planners* 27 (1961): 134–140.

Gans, H. J. *The Levittowners*. New York: Pantheon Books, 1967.

Gans, H. J. *People and Plans*. New York: Basic Books, 1968.

Garvey, J., Jr. "New, Expanding and Renewed Town Concepts." *Assessors Journal* 4 (1969): 49–57.

Gibberd, F. *Town Design*. London: Architectural Press, 1953.

Glazer, N., and Moynihan, D. P. *Beyond the Melting Pot*. Cambridge: M.I.T. Press, 1963.

Godschalk, D. R. "Comparative New Community Design." *Journal of the American Institute of Planners* 33 (1967): 371–387.

Goetze, R. *Building Neighborhood Confidence: A Humanistic Strategy for Urban Housing*. Cambridge: Ballinger Publishing Company, 1976.

Gold, R. "Urban Violence and Contemporary Defensive Cities." *Journal of the American Institute of Planners* 36 (1970): 146–159.

Gordon, N. J. "China and the Neighborhood Unit." *The American City* 61 (1946): 112–113.

Goss, A. "Neighborhood Units in British New Towns." *Town Planning Review* 32 (1961): 66–82.

Greenbie, B. B. "Social Territory, Community Health and Urban Planning." *Journal of the American Institute of Planners* 40 (1974): 74–82.

Gropius, W. *Rebuilding Our Communities*. Chicago: Paul Theobald, 1945.

Gutnov, A., Babunor, A., Djumenton, G., Kharitonva, S., Lezava, I., and Sadovskij, S. *The Ideal Communist City*. New York: George Braziller, 1968.

Guttenberg, A. Z. "City Encounter and 'Desert' Encounter: Two Sources of American Regional Planning Thought." *Journal of the American Institute of Planners* 44 (1978): 399–411.

Hartman, C. "Comment on 'Neighborhood Revitalization and Displacement: A Review of the Evidence.'" *Journal of the American Planning Association* 45 (1979): 488–491.

Harvey, D. *Social Justice and the City*. Baltimore: Johns Hopkins University Press, 1973.

Hendricks, F. and MacNair, M. Concepts of Environment Quality Standards Based on Life Styles with Special Emphasis on Family Cycle. In *Final Report on Planning, Designing and Managing the Residential Environment: Stage One*, I. Robinson (ed.). Los Angeles: School of Urban and Regional Planning, University of Southern California, 1969.

Herbert, G. "The Neighborhood Unit Principle and Organic Theory." *The Sociological Review* 11 (1963): 165–213.

Horton, F. E., and Reynolds, D. R. "Effects of the Urban Spatial Structure on Individual Behavior." *Economic Geography* 47 (1971): 36–48.

Isaacs, R. "Are Urban Neighborhoods Possible?" *Journal of Housing* 5 (1948): 177–180. (a)

Isaacs, R. "The 'Neighborhood Unit' is an Instrument of Segregation." *Journal of Housing* 5 (1948): 215–219. (b)

Isaacs, R. "The Neighborhood Theory." *Journal of the American Institute of Planners* 14 (2) (1948): 15–23. (c)

Isaacs, R. "The Neighborhood Concept in Theory and Application." *Land Economics* 25 (1949): 73–81.

Keller, S. *The Urban Neighborhood: A Sociological Perspective*. New York: Random House, 1968.

Kuhn, T. S. *The Structure of Scientific Revolutions*. Chicago: Chicago University Press, 1970 (second and enlarged ed.).

Kuper, L. "Social Science Research and the Planning of Urban Neighborhoods." *Social Forces* 29 (1951): 241–247.

Lansing, J. B., Marans, R. W., and Zehner, R. B. *Planned Residential Environments*. Ann Arbor, Mich,: Survey Research Center, 1970.

Lee, T. "Urban Neighborhood as a Socio-spatial Schema." *Human Relations* 21 (1968): 241–288.

Lefebvre, H. "The Neighborhood and Neighborhood Life." *Planification Habitat Information*. 75 (1973): 3–8.

Levy, F. S., Meltsner, A. J., and Wildavsky, A. *Urban Outcomes: Schools, Streets, and Libraries*. Berkeley: University of California Press, 1974.

Lineberry, R. L. "Equality, Public Policy and Public Services: The Underclass Hypothesis and the Limits to Equality." *Politics and Policy* 4 (1975): 67–84.

London County Council. *The Planning of a New Town: Data and Design Based on a Study for a New Town at Hook, Hampshire*. London: Author, 1961.

Low, N. "Centrism and the Provision of Services in Residential Areas." *Urban Studies* 12 (1975): 177–191.

Lowenthal, D. *Environmental Assessment: A Comparative Analysis of Four Cities*. New York: American Geographical Society, 1972.

Lubove, R. *The Progressives and the Slums*. Pittsburgh: University of Pittsburgh Press, 1962.

Lynch, K. *The Image of the City*. Cambridge: M.I.T. Press, 1960.

Lynch, K. *Managing the Sense of the Region*. Cambridge: M.I.T. Press, 1976.

Lynch, K. *A Theory of Good City Form*. Cambridge: M.I.T. Press, 1981.

Mann, P. H. "The Socially Balanced Neighborhood Unit." *Town Planning Review* 29 (1958): 91–98.

Maslow, A. H. *Motivation and Personality* (2nd ed.), New York: Harper & Row, 1970.

Mawby, R. I. "Defensible Space: A Theoretical and Empirical Appraisal." *Urban Studies,* 14, (1977): 169–180.

McKie, R. "Cellular Renewal: A Policy for the Older Housing Areas." *Town Planning Review* 45, (1974): 274–290.

Merry, S. "Defensible Space Undefended: Social Factors in Crime Control through Environmental Design." *Urban Affairs Quarterly*. 16 (1981): 397–422.

Michelson, W. "Urban Sociology as an Aid to Urban Physical Development: Some Research Strategies." *Journal of the American Institute of Planners* 34 (1968): 105–108.

Michelson, W. *Man and His Urban Environment: A Sociological Approach*. Reading, Mass.: Addison-Wesley, 1970.

Milgram, S. "The Experience of Living in Cities." *Science* 167 (1970): 1461–1468.

Milton Keynes Development Corporation. *The Plan for Milton Keynes*. Vol. 2, Wavendon near Bletchley, Buckinghamshire. England: Milton Keynes, 1970.

Mumford, L. "Introduction." In *Toward New Towns for American,* edited by C. S. Stein. Cambridge: M.I.T. Press, 1951.

Mumford, L. "The Neighborhood and the Neighborhood Unit." *Town Planning Review* 24 (1954): 256–270.

Mumford, L. *The City in History: Its Origins, Its Transformations, and Its Prospects.* New York: Harcourt, Brace and World, 1961.

Neiman, M. "From Plato's Philosopher King to Bish's Tough Purchasing Agent." *Journal of the American Institute of Planners* 41 (1975): 66–72.

Newman, O. *Defensible Space: Crime Prevention through Urban Design.* New York: Collier, 1972.

Osborn, F. J. and Whittick, A. *The New Towns: The Answer to Megalopolis?* London: Leonard Hill Books, 1969.

Pahl, R. E. *Patterns of Urban Life.* London: Longmans, Green, 1970.

Park, R. *Human Communities.* New York: Free Press, 1952.

Perin, C. *With Man in Mind: An Interdisciplinary Prospectus for Environmental Design.* Cambridge: M.I.T. Press, 1970.

Perry, C. A. *Housing for the Machine Age.* New York: Russell Sage Foundation, 1939.

Pierce, S. *Analysis Across Several Population Variables of Neighborhood Elements Desired in the Residential Area.* Directed Research, Graduate Program of Urban and Regional Planning, University of Southern California, 1976.

Polanyi, M. *Personal Knowledge: Towards a Post-Critical Philosophy.* Chicago: University of Chicago Press, 1958.

Porteous, J. D. *Environment and Behavior: Planning and Everyday Urban Life.* Reading, Mass.: Addison-Wesley, 1977.

Protzen, J. P. "The Poverty of the Pattern Language." *UC Berkeley Newsletter* 1 (1977): 2–4, 15.

Rainwater, L. "Fear and the House-as-Haven in the Lower Class." *Journal of the American Institute of Planners* 32 (1966): 23–31.

Rawls, J. *A Theory of Justice.* Cambridge, Mass.: Belknap Press, 1971.

Reimer, S. "The Neighborhood Concept in Theory and Application." *Land Economics* 25 (1949): 69–72.

Reimer, S. "Hidden Dimensions of Neighborhood Planning." *Land Economics* 26 (1950): 197–201.

Rich, R. C. "Equity and Institutional Design in Urban Service Delivery." *Urban Affairs Quarterly* 12 (1977): 383–410.

Rich, R. C. "Neglected Issues in the Study of Urban Service Distributions: a Research Agenda." *Urban Studies* 16 (1979): 143–156.

"Richard Sennett Lectures on Democratic Theory and Urban Form." *HGSD News.* Vol. 10. Cambridge: Harvard Graduate School of Design, 1982.

Rittel, H., and Webber, M. "Dilemmas in a General Theory of Planning." *Policy Sciences* 4 (1973): 155–169.

Robinson, I. M., Baer, W. C., Banerjee, T. K., and Flachsbart, P. G. "Trade-Off Games." In *Behavioral Research Methods in Environmental Design,* edited by W. Michelson. Stroudsburg, Pa.: Dowden, Hutchinson and Ross, 1975.

Salley, M. A. "Public Transportation and the Needs of New Communities." *Traffic Quarterly* 16 (1972): 33–49.

Schon, D. *The Reflective Practitioner.* New York: Basic Books, 1982.

Schulze, W. D. "Ethics, Economics and the Value of Safety." In *Societal Risk Assessment: How Safe Is Enough?* edited by R. S. Schwing and W. A. Albers, Jr. New York: Plenum Press, 1980.

Sennett, R. "The Brutality of Modern Families." *Transaction* 7 (1970): 29–37. (a)

Sennett, R. *The Uses of Disorder.* New York: Vintage Books, 1970. (b)

Sims, W. R. *Neighborhoods: Columbus Neighborhood Definition Study.* Department of Development, City of Columbus, Ohio, 1973.

Slidell, J. B. *The Shape of Things to Come? An Evaluation of the Neighborhood Unit as an Organizing Schema for American New Towns.* Chapel Hill: Center for Urban and Regional Studies, University of North Carolina, 1972.

Smithson, A., and Smithson, P. *Urban Structuring.* New York: Reinhold, 1967.

Solow, A. A., Ham, C. E., and Donnelly, E. O. "The Concept of Neighborhood Unit: Its Emergence and Influence on Residential Environment Planning and Development." In *Final Report on Planning, Designing and Managing the Residential Envrionment: Stage One,* edited by I. M. Robinson. Los Angeles: Graduate Program of Urban and Regional Planning, University of Southern California, 1969.

Spivack, M. "Archtypal Place." In *Environmental Design Research. Vol. 1,* edited by W. F. E. Preiser. Stroudsburg, Pa.: Dowden, Hutchinson and Ross, 1973.

Stein, C. S. *Toward New Towns for America.* Cambridge: M.I.T. Press, 1957.

Sumka, H. J. "Neighborhood Revitalization and Displacement: A Review of the Evidence." *Journal of the American Planning Association* 45 (1979): 480–487.

Suttles, G. D. *The Social Construction of Communities.* Chicago: The University of Chicago Press, 1973.

Suttles, G. "Community Design: The Search for Participation in a Metropolitan Society." In *Metropolitan America in Contemporary Perspective,* edited by A. H. Hawley and V. P. Rock. New York: Wiley, 1975.

Tannenbaum, J. "The Neighborhood: A Socio-psychological Analysis." *Land Economics* 24 (1948): 358–369.

Thullier, R. *Air Quality Considerations in Residential Planning.* Vol. 1. Menlo Park, Calif.: S.R.I. International, 1978.

Toffler, A. *Future Shock.* New York: Bantam, 1970.

U.S. Department of Housing and Urban Development. *The 1978 HUD Survey of the Quality of Community Life: A Data Book.* Washington, D.C.: Office of the Policy Development and Research, HUD, n.d.

Ward, C. *Vandalism.* London: Architectural Press, 1973.

Webber, M. M. "Order in diversity: Community without Propinquity." In *Cities and Space: The Future Use of Urban Land,* edited by L. Wingo. Baltimore: Johns Hopkins University Press, 1963.

Webber, M. M. "The Urban Place and the Nonplace Urban Realm." In *Explorations into Urban Structure,* edited by M. M. Webber. Philadelphia: University of Pennsylvania Press, 1964.

Webber, M. M., and Webber, C. C. "Culture, Territoriality, and the Elastic Mile." In *Taming Megalopolis.* Vol. 1, edited by H. W. Eldredge. New York: Anchor Books, 1967.

Wehrly, M. S. "Comment on the Neighborhood Theory." *Journal of the American Institute of Planners* 14 (1948): 32–34.

Werthman, C., Mandell, J. S., and Dienstfrey, T. *Planning and the Purchase Decision: Why People Buy in Planned Communities?* Berkeley: Institute of Urban and Regional Development, University of California, 1965.

White, M., and White, L. *The Intellectual versus the City.* New York: Mentor Books, 1962.

Willis, M. "Sociological Aspects of Urban Structure: Comparison of Residential Groupings Proposed in Planning New Towns." *Town Planning Review* 39 (1969): 296–306.

Willmott, P. "Housing Density and Town Design in a New Town." *Town Planning Review* 33 (1962): 114–127.

Willmott, P. "Social Research and New Communities." *Journal of the American Institute of Planners* 32 (1967): 387–398.

Wohlwill, J. *A Psychologist Looks at Land Use.* Paper presented at the Symposium on "Psychology and Environment in the 1980's," held at the University of Mississippi, Columbia, 1975.

Wolch, J. "Residential Location and the Provision of Human Services." *Professional Geographer* 31 (1979): 271–276.

Wolpert, J. "Behavioral Aspects of the Decision to Migrate." *Papers and Proceedings of the Regional Science Association* 15 (1965): 159–169.

Appendix I

Survey Questionnaire

RESEARCH ON THE RESIDENTIAL ENVIRONMENT
BACKGROUND QUESTIONS

CONFIDENTIAL

INSTRUCTIONS

1. This questionnaire takes just a few minutes to complete. Please take your time and read the questions carefully before answering. *1–6/*
2. The questionnaire is in two parts. Part I refers to the entire household as a unit. It should be answered by the head of the household. Part II deals with the household member who was selected for the interview, and should be answered by that person. If the head of the household was the person selected for the interview he/she should answer both Part I and Part II, skipping questions 2 - 4 in Part II.
3. The numbers in the right hand margins are for office use, to assist in machine processing. Please disregard them as you fill out the questionnaire.
4. If possible, complete this questionnaire before the date set for your appointment with the interviewer. If, however, you have difficulty with any of the questions please feel free to ask the interviewer to help you when he or she arrives.

PART I

HOUSEHOLD QUESTIONS

To be answered by the head of the household.

1. How many persons live in this household? Please remember to include new babies or someone who usually lives here but is away right now.

______________ *7,8/*
Number of persons

2. How many of these people are children under 17? ______________ *9,10/*

3. What is the age and sex of each child under 17?

	AGE	SEX *(Circle one)*	
1.	______	M F	*11–13/*
2.	______	M F	*14–16/*
3.	______	M F	*17–19/*
4.	______	M F	*20–22/*
5.	______	M F	*23–25/*
6.	______	M F	*26–28/*
7.	______	M F	*29–31/*
8.	______	M F	*32–34/*
9.	______	M F	*35–37/*
10.	______	M F	*38–40/*

4. What is the sex of the head of the household? ☐ M ☐ F *41/*

5. What is the present employment status of the head of household?

(CIRCLE ONE) *42/*

Working full time 1
Working part time 2
Unemployed 3
Retired, at leisure 4
Keeping house 5
In school 6
Other *(Please specify)* . . . . 7 ______________________

(Note: Items 11, 12, 13C, 13D, 18 and 19 were not included in the truncated version of the survey questionnaire used for the lower-income black and Hispanic groups.)

CONFIDENTIAL

6. If employed:

a. What kind of work does the head of household do? *(EXAMPLE: teacher, sales clerk, foreman)*

____

____ 43–45/

b. In what type of industry is the head of household employed? *(EXAMPLE: high school, shoe store, auto assembly plant)* ____

____ 46–48/

7. If this is a husband and wife household, does the wife work half time or more? _____ Yes _____ No 49/

8. What is the highest level of education completed by the head of household?
(CIRCLE HIGHEST GRADE COMPLETED)

NONE	ELEMENTARY	HIGH SCHOOL	COLLEGE	POST GRADUATE
0	1 2 3 4 5 6 7 8	9 10 11 12	13 14 15 16	17 18 19 20+

50,51/

Other schooling *(EXAMPLE: trade-technical, business school)*

52/

____ *Type of School* _______________ *Years Completed* 53/

9. Is this home:

(CIRCLE ONE)

A single family, detached house 1 54/
A duplex 2
A triplex, fourplex 3
A row house (townhouse) 4
A garden apartment or condominium 5
A low rise elevator apartment or condominium
(3-7 floors) 6
A high-rise apartment or condominium
(8+ floors) 7
A mobile home 8
Other *(Please specify)* . . . . 9 _______________

10. Is this home:

(CIRCLE ONE)

a. Owned or being bought by a member of this household . . . 1 55/
b. Rented 2
c. Other *(Please specify)* . . . . 3 _______________

11. If you live in a home which you own or are purchasing, what is the present value of the home? In other words, what would it bring on the market today?

(CIRCLE ONE)

Less than $10,000 . . . . 1 56,57/
$10,000 – $12,499 2
$12,500 – $14,999 3
$15,000 – $17,499 4
$17,500 – $19,999 5
$20,000 – $24,999 6
$25,000 – $34,999 7
$35,000 – $49,999 8
$50,000 – $59,999 9
$60,000 – $74,999 10
$75,000 and over 11
Don't know 12

CONFIDENTIAL

12. If you are renting, what is the monthly rent?

(CIRCLE ONE)

Less than $60.00	1	58,59/
$ 60.00 – $ 79.99	2	
$ 80.00 – $ 99.99	3	
$100.00 – $119.99	4	
$120.00 – $149.99	5	
$150.00 – $199.99	6	
$200.00 – $299.99	7	
$300.00 – $399.99	8	
$400.00 – $499.99	9	
$500.00 – $599.99	10	
$600.00 and over	11	

13. What was the family income of this household last year, 1971? By family income we mean the combined income from all sources of all family members, before taxes. However, any portion of this income which goes for the support of another household should not be included.

(CIRCLE ONE)

Under $2,000	1	60,61/
$ 2,000 – $ 3,999	2	
$ 4,000 – $ 5,999	3	
$ 6,000 – $ 7,999	4	
$ 8,000 – $ 9,999	5	
$10,000 – $11,999	6	
$12,000 – $14,999	7	
$15,000 – $19,999	8	
$20,000 – $24,999	9	
$25,000 – $29,999	10	
$30,000 – $39,999	11	
$40,000 – $49,999	12	
$50,000 and over	13	

14. How many people were dependent on that income last year (1971)? ____________ 62,63/

Number

15. How many persons in the household were employed half time or more? ____________ 64/

Number

16. How many vehicles are owned by members of this household?

(FILL IN NUMBER)

Automobiles	________	65/
Trucks, campers, vans, etc.	________	66/
Motorcycles, motorbikes	________	67/
Bicycles	________	68/
Other *(Please describe)*	________	69/

17. What other form of transportation is available to household members?

Bus	1	70/
Car pool	2	
Friend's car	3	
Other *(Please describe)*	4 ________	

18. How many minutes away (walking time) is the nearest bus stop? ____________ 71,72/

Don't know ____________ 73,74/

75–78/0101

CONFIDENTIAL

PART II

RESPONDENT QUESTIONS

These questions should be answered by the person selected for the interview. If that person is the head of the household he/she may skip questions 2 – 4 in Part II. 1–6/

1. What is your relationship to the head of the household?

(CIRCLE ONE)

Head of the household 1 7/
Spouse of the head 2
Son or daughter of the head or spouse 3
Father or mother of the head or spouse 4
Other relative of the head or spouse 5
Roomers, boarders 6
Others, not related to the head *(Please specify)*. . 7 ____________

2. What is your present employment status?

Working full time 1 8/
Working part time 2
Unemployed 3
Retired, at leisure 4
Keeping house 5
In school 6
Other *(Please specify)* . . . . 7 ____________

3. If employed:

a. What kind of work do you do? *(EXAMPLE: teacher, sales clerk, foreman)*

____

____________ 9–11

b. In what type of industry are you employed? *(EXAMPLE: high school, shoe store, auto assembly plant)* ____________

____________ 12–14/

4. What is the highest level of education you have completed? *(CIRCLE HIGHEST GRADE COMPLETED)*

NONE	ELEMENTARY	HIGH SCHOOL	COLLEGE	POST GRADUATE	
0	1 2 3 4 5 6 7 8	9 10 11 12	13 14 15 16	17 18 19 20+	15,16/

Other schooling *(EXAMPLE: trade-technical, business school)* ____________

____________ 17/

Type of School *Years Completed* 18/

CONFIDENTIAL

5. There are lots of different types of Americans. In this study we are especially interested in Blacks, Whites, and Mexican Americans because these groups account for most of the people in Los Angeles. What do you consider your race or ethnic background to be?

(CIRCLE ONE)

Black or Negro 1
White or Caucasian 2
Brown or Mexican American 3
Other *(Please specify)* . . . . 4 ______________________ *19/*

6. We would like you to indicate the size of the area in which you lived for each 5 year period of your life *up to and including the present.* **Do this by putting the correct letter from Column I ("City Type") on each line in Column II ("Age"). For example if you lived in Los Angeles between the ages of 16 – 20, put the letter A on line 4. If you lived in cities of different sizes during any one 5 year period, you may put as many letters as necessary on that line.**

Column I *CITY TYPE*	Column II *AGE*		
A. Big City	1. 0 – 5	________	*20–22/*
B. Medium Sized City	2. 6 – 10	________	*23–25/*
C. Suburb	3. 11 – 15	________	*26–28/*
D. Small Town	4. 16 – 20	________	*29–31/*
E. Rural	5. 21 – 25	________	*32–34/*
	6. 26 – 30	________	*35–37/*
	7. 31 – 35	________	*38–40/*
	8. 36 – 40	________	*41–43/*
	9. 41 – 45	________	*44–46/*
	10. 46 – 50	________	*47–49/*
	11. 51 – 55	________	*50–52/*
	12. 56 – 60	________	*53–55/*
	13. 61 – 65	________	*56–58/*
	14. Over 65	________	*59–61/*

7. What was your individual income from all sources, before taxes, last year, 1971?

(CIRCLE ONE)

Under $2,000 . . . . 1 *62,63/*
$ 2,000 – $ 3,999 2
$ 4,000 – $ 5,999 3
$ 6,000 – $ 7,999 4
$ 8,000 – $ 9,999 5
$10,000 – $11,999 6
$12,000 – $14,999 7
$15,000 – $19,999 8
$20,000 – $24,999 9
$25,000 – $29,999 10
$30,000 – $39,999 11
$40,000 – $49,999 12
$50,000 and over 13

75–78/0102

RESEARCH ON THE RESIDENTIAL ENVIRONMENT

UNIVERSITY OF SOUTHERN CALIFORNIA

Stage Two Survey Interview

Area Name ____________________

Interviewer's Code No. ____________________

Interview No. ____________________

Date of Interview ____________________

Time Started ____________________

Time Finished ____________________

Length of Interview (min.) ____________________

1. Please describe the residential area you live in.
(PROBE: What else?)

1-6/

7,8/

9,10/

11,12/

13,14/

15,16/

17,18/

19,20/

21,22/

---

2. In what ways is your residential area a good place to live?
(PROBE: What else? or What other ways?)

23,24/

25,26/

27,28/

29,30/

31,32/

33,34/

35,36/

37,38/

---

3. In what ways is your residential area not a good place to live? (PROBE: What else? or What other ways?)

39,40/
41,42/
43,44/
45,46/
47,48/
49,50/
51,52/
53,54/

---

4. All things considered is your residential area ... (CIRCLE NUMBER)

excellent 1
good 2
average 3 55/
fair, or 4
poor 5
as a place to live?

---

5. The next question you fill out yourself. The instructions are on the first page. HAND RESPONDENT THE SHEETS MARKED QUESTION # 5.

QUESTION #5

INSTRUCTIONS

We want to know how you picture your residential area. On the next two pages are listed many pairs of words which have opposite meanings. Each pair of words is separated by a line in which the meaning of each location on the line is as follows:

noisy	______ :	______ :	______ :	______ :	______ :	______ :	______	quiet
	1	2	3	4	5	6	7	
	very	quite	slightly	neither	slightly	quite	very	

For each pair of words on the next two pages check the location that best describes your residential area. For example, if your area is "slightly quiet" you would have checked location #5. On the other hand, if your area is "quite noisy" you would have checked location #2. Repeat this task for each pair of words on the last two pages, only this time check the location that best describes your <u>ideal</u> residential area. Do not spend too much time in deciding. Your <u>first impression</u> is what we would like.

My residential area is . . .

	1 very	2 quite	3 slightly	4 neither	5 slightly	6 quite	7 very		
Noisy	_____ :	_____ :	_____ :	_____ :	_____ :	_____ :	_____	Quiet	7/
Friendly	_____ :	_____ :	_____ :	_____ :	_____ :	_____ :	_____	Hostile	8/
Young	_____ :	_____ :	_____ :	_____ :	_____ :	_____ :	_____	Old	9/
Poor	_____ :	_____ :	_____ :	_____ :	_____ :	_____ :	_____	Rich	10/
Unsafe	_____ :	_____ :	_____ :	_____ :	_____ :	_____ :	_____	Safe	11/
High Status	_____ :	_____ :	_____ :	_____ :	_____ :	_____ :	_____	Low Status	12/
Fancy	_____ :	_____ :	_____ :	_____ :	_____ :	_____ :	_____	Plain	13/
Delicate	_____ :	_____ :	_____ :	_____ :	_____ :	_____ :	_____	Rugged	14/
Ordered	_____ :	_____ :	_____ :	_____ :	_____ :	_____ :	_____	Chaotic	15/
Personal	_____ :	_____ :	_____ :	_____ :	_____ :	_____ :	_____	Impersonal	16/
Rough	_____ :	_____ :	_____ :	_____ :	_____ :	_____ :	_____	Smooth	17/
Relaxed	_____ :	_____ :	_____ :	_____ :	_____ :	_____ :	_____	Tense	18/
Simple	_____ :	_____ :	_____ :	_____ :	_____ :	_____ :	_____	Complex	19/
Worthless	_____ :	_____ :	_____ :	_____ :	_____ :	_____ :	_____	Valuable	20/
Healthy	_____ :	_____ :	_____ :	_____ :	_____ :	_____ :	_____	Sick	21/
Private	_____ :	_____ :	_____ :	_____ :	_____ :	_____ :	_____	Public	22/
Active	_____ :	_____ :	_____ :	_____ :	_____ :	_____ :	_____	Passive	23/
Unique	_____ :	_____ :	_____ :	_____ :	_____ :	_____ :	_____	Common	24/
Exciting	_____ :	_____ :	_____ :	_____ :	_____ :	_____ :	_____	Dull	25/
Modern	_____ :	_____ :	_____ :	_____ :	_____ :	_____ :	_____	Traditional	26/
Inconvenient	_____ :	_____ :	_____ :	_____ :	_____ :	_____ :	_____	Convenient	27/
	1	2	3	4	5	6	7		

My residential area is . . .

	1 very	2 quite	3 slightly	4 neither	5 slightly	6 quite	7 very		
Crowded	____ :	____ :	____ :	____ :	____ :	____ :	____	Spacious	28/
Beautiful	____ :	____ :	____ :	____ :	____ :	____ :	____	Ugly	29/
Comfortable	____ :	____ :	____ :	____ :	____ :	____ :	____	Uncomfortable	30/
Satisfying	____ :	____ :	____ :	____ :	____ :	____ :	____	Unsatisfying	31/
Integrated	____ :	____ :	____ :	____ :	____ :	____ :	____	Segregated	32/
Austere	____ :	____ :	____ :	____ :	____ :	____ :	____	Sensuous	33/
Formal	____ :	____ :	____ :	____ :	____ :	____ :	____	Casual	34/
Changing	____ :	____ :	____ :	____ :	____ :	____ :	____	Lasting	35/
Man-made	____ :	____ :	____ :	____ :	____ :	____ :	____	Natural	36/
Secure	____ :	____ :	____ :	____ :	____ :	____ :	____	Insecure	37/
Dirty	____ :	____ :	____ :	____ :	____ :	____ :	____	Clean	38/
Warm	____ :	____ :	____ :	____ :	____ :	____ :	____	Cool	39/
Neglected	____ :	____ :	____ :	____ :	____ :	____ :	____	Cared-for	40/
Calm	____ :	____ :	____ :	____ :	____ :	____ :	____	Restless	41/
Small	____ :	____ :	____ :	____ :	____ :	____ :	____	Large	42/
Useless	____ :	____ :	____ :	____ :	____ :	____ :	____	Useful	43/
Aloof	____ :	____ :	____ :	____ :	____ :	____ :	____	Talkative	44/
Colorful	____ :	____ :	____ :	____ :	____ :	____ :	____	Colorless	45/
Fast	____ :	____ :	____ :	____ :	____ :	____ :	____	Slow	46/
Unfamiliar	____ :	____ :	____ :	____ :	____ :	____ :	____	Familiar	47/
Rigid	____ :	____ :	____ :	____ :	____ :	____ :	____	Flexible	48/
	1	2	3	4	5	6	7		75-78/050

My <u>ideal</u> residential area is

	1 very	2 quite	3 slightly	4 neither	5 slightly	6 quite	7 very		
Noisy	_____	_____	_____	_____	_____	_____	_____	Quiet	7/
Friendly	_____	_____	_____	_____	_____	_____	_____	Hostile	8/
Young	_____	_____	_____	_____	_____	_____	_____	Old	9/
Poor	_____	_____	_____	_____	_____	_____	_____	Rich	10/
Unsafe	_____	_____	_____	_____	_____	_____	_____	Safe	11/
High Status	_____	_____	_____	_____	_____	_____	_____	Low Status	12/
Fancy	_____	_____	_____	_____	_____	_____	_____	Plain	13/
Delicate	_____	_____	_____	_____	_____	_____	_____	Rugged	14/
Ordered	_____	_____	_____	_____	_____	_____	_____	Chaotic	15/
Personal	_____	_____	_____	_____	_____	_____	_____	Impersonal	16/
Rough	_____	_____	_____	_____	_____	_____	_____	Smooth	17/
Relaxed	_____	_____	_____	_____	_____	_____	_____	Tense	18/
Simple	_____	_____	_____	_____	_____	_____	_____	Complex	19/
Worthless	_____	_____	_____	_____	_____	_____	_____	Valuable	20/
Healthy	_____	_____	_____	_____	_____	_____	_____	Sick	21/
Private	_____	_____	_____	_____	_____	_____	_____	Public	22/
Active	_____	_____	_____	_____	_____	_____	_____	Passive	23/
Unique	_____	_____	_____	_____	_____	_____	_____	Common	24/
Exciting	_____	_____	_____	_____	_____	_____	_____	Dull	25/
Modern	_____	_____	_____	_____	_____	_____	_____	Traditional	26/
Inconvenient	_____	_____	_____	_____	_____	_____	_____	Convenient	27/
	1	2	3	4	5	6	7		

My <u>ideal</u> residential area is

	1 very	2 quite	3 slightly	4 neither	5 slightly	6 quite	7 very		
Crowded	____ :	____ :	____ :	____ :	____ :	____ :	____	Spacious	28/
Beautiful	____ :	____ :	____ :	____ :	____ :	____ :	____	Ugly	29/
Comfortable	____ :	____ :	____ :	____ :	____ :	____ :	____	Uncomfortable	30/
Satisfying	____ :	____ :	____ :	____ :	____ :	____ :	____	Unsatisfying	31/
Integrated	____ :	____ :	____ :	____ :	____ :	____ :	____	Segregated	32/
Austere	____ :	____ :	____ :	____ :	____ :	____ :	____	Sensuous	33/
Formal	____ :	____ :	____ :	____ :	____ :	____ :	____	Casual	34/
Changing	____ :	____ :	____ :	____ :	____ :	____ :	____	Lasting	35/
Man-made	____ :	____ :	____ :	____ :	____ :	____ :	____	Natural	36/
Secure	____ :	____ :	____ :	____ :	____ :	____ :	____	Insecure	37/
Dirty	____ :	____ :	____ :	____ :	____ :	____ :	____	Clean	38/
Warm	____ :	____ :	____ :	____ :	____ :	____ :	____	Cool	39/
Neglected	____ :	____ :	____ :	____ :	____ :	____ :	____	Cared-for	40/
Calm	____ :	____ :	____ :	____ :	____ :	____ :	____	Restless	41/
Small	____ :	____ :	____ :	____ :	____ :	____ :	____	Large	42/
Useless	____ :	____ :	____ :	____ :	____ :	____ :	____	Useful	43/
Aloof	____ :	____ :	____ :	____ :	____ :	____ :	____	Talkative	44/
Colorful	____ :	____ :	____ :	____ :	____ :	____ :	____	Colorless	45/
Fast	____ :	____ :	____ :	____ :	____ :	____ :	____	Slow	46/
Unfamiliar	____ :	____ :	____ :	____ :	____ :	____ :	____	Familiar	47/
Rigid	____ :	____ :	____ :	____ :	____ :	____ :	____	Flexible	48/
	1	2	3	4	5	6	7		75-78/0502

6. PLACE THE FOLLOWING YELLOW CARDS BEFORE THE RESPONDENT:

1. THINGS ACTUALLY IN MY AREA

2. THINGS ACTUALLY OUT OF MY AREA

3. DON'T KNOW

HAND RESPONDENT A DECK OF GREEN CARDS.

On each of those cards are printed different things which may or may not be in the area which you described to me a moment ago. As you read each card decide which of these categories the card best fits under, and then place the card there.

UPON COMPLETION OF THE TASK PLACE EACH LABEL CARD ON ITS RESPECTIVE PILE AND WRAP IT WITH A RUBBER BAND.

---

7. PLACE THE FOLLOWING YELLOW CARDS BEFORE THE RESPONDENT:

1. THINGS I WANT IN MY AREA

2. THINGS I WANT OUT OF MY AREA

3. EITHER IN OR OUT IS FINE

HAND RESPONDENT THE OTHER DECK OF GREEN CARDS.

Those green cards have the same things printed on them as the ones you used before. This time we are interested in the things you may or may not <u>want</u> in your area. As you read each card decide which of these categories the card best fits under, and then place the card there.

UPON COMPLETION OF THE TASK PLACE EACH LABEL CARD ON ITS RESPECTIVE PILE AND WRAP IT WITH A RUBBER BAND.

---

8. HAND RESPONDENT A PENCIL AND THE SHEET OF PAPER MARKED "MAP OF MY RESIDENTIAL AREA". LEAVE THIS SHEET OF PAPER BEFORE THE RESPONDENT UNTIL QUESTION 10 IS COMPLETED.

On this sheet of paper draw a map of your residential area. Put as many details as you can on the map. (PROBE IF NECESSARY UPON COMPLETION OF THE TASK: Mark your home on the map AND label the things you put on your map.)

A. How would you go to get to the farthest edge of this map? Would you...

Walk	1	56/
Take an automobile	2	
Take a bicycle	3	
Take a motorcycle................	4	
Take a bus	5	

B. How long would it take you in minutes to get to the farthest edge of the map? ______________________ 57,58/

---

9. How important is it for you to live in a place that you consider a neighborhood? Is it... (CIRCLE NUMBER)

very important	fairly important	not important at all	
1	2	3	59,60/

PROBE: Why do you say that? 61,62/

63,64/

65,66/

67,68/

69,70/

71,72/

---

10. With a "Yes" or "No" answer, can you identify your neighborhood on this map? (CIRCLE NUMBER)

Yes........................	1	73/
No (SKIP TO Q.11)............	2	

A. (IF YES) With respect to the size of your residential area as shown on this map, is your neighborhood: (CIRCLE NUMBER)

smaller	(ASK Q.B).....................	1	74/
the same size, or	(SKIP TO Q.11).................	2	
larger..................	(ASK Q.B)....................	3	

B. (IF NUMBERS 1 OR 3 ARE CIRCLED) Please mark the area on the map that you consider your neighborhood and, if possible, indicate its boundaries. (UPON COMPLETION OF TASK, REMOVE MAP)

75-78/0201

11. The following questions pertain to your residential area. Please answer "**Yes**" or "No" as I read each question.

A. Is it an area known by a certain name? (CIRCLE NUMBER) 1-6/

Yes 1 7/

No (SKIP TO Q.B.) 2

(IF YES) What is that name? ______________________ 8,9/

B. Is it an area bounded by certain streets? (CIRCLE NUMBER)

Yes 1 10/

No (SKIP TO Q.C.) 2

(IF YES) What are those streets?

(a) ______________________ 11/

(b) ______________________ 12/

(c) ______________________ 13/

(d) ______________________ 14/

(e) ______________________ 15/

C. Is it an area which has certain natural or man-made features, landmarks, or other characteristics? (CIRCLE NUMBER)

Yes 1 16/

No (SKIP TO Q.D.)....... 2

(IF YES) What are they? (PROBE: What else?)

(a) ______________________ 17/

(b) ______________________ 18/

(c) ______________________ 19/

(d) ______________________ 20/

(e) ______________________ 21/

D. Is it an area served by certain schools? (CIRCLE NUMBER)

Yes .... 1 — 22/

No (SKIP TO Q.E.)..... 2

(IF YES) What type of schools are they? (CHECK)

(a) Elementary ____ — 23/

(b) Junior High ____ — 24/

(c) Senior High ____ — 25/

(d) College or University ____ — 26/

(e) Other ____ (SPECIFY: ________________) — 27/

E. Is it an area which has a certain shopping area? (CIRCLE NUMBER)

Yes .... 1 — 28/

No (SKIP TO Q.F.)...... 2

(IF YES) What is the shopping area's name?

______________________________ — 29,30/

F. Is it an area which has a certain kind of people? (CIRCLE NUMBER)

Yes .... 1 — 31/

No (SKIP TO Q.G.)...... 2

(IF YES) What kind of people are they? (PROBE: What else about these people?)

(a) ______________________________ — 32,33/

(b) ______________________________ — 34,35/

(c) ______________________________ — 36,37/

(d) ______________________________ — 38,39/

(e) ______________________________ — 40,41/

G. Is it an area which has a certain kind of housing? (CIRCLE NUMBER)

Yes .... 1 — 42/

No (SKIP TO Q.H.).......... 2

(IF YES) What kind of housing is it? (PROBE: What else about this housing?)

(a) ______________________ 43,44/

(b) ______________________ 45,46/

(c) ______________________ 47,48/

(d) ______________________ 49,50/

(e) ______________________ 51,52/

H. Is it an area where people share common concerns or problems? (CIRCLE NUMBER)

Yes .... 1 — 53/

No (SKIP TO Q.I.).......... 2

(IF YES) What are those concerns or problems? (PROBE: What else?)

(a) ______________________ 54,55/

(b) ______________________ 56,57/

(c) ______________________ 58,59/

(d) ______________________ 60,61/

(e) ______________________ 62,63/

I. Is there something else that I haven't mentioned that this area has? (CIRCLE NUMBER)

Yes 1

64/

No (SKIP TO Q.12)...... 2

(IF YES) What is it? (PROBE: What else?)

(a) ______________________________ 65,66/

(b) ______________________________ 67,68/

(c) ______________________________ 69,70/

(d) ______________________________ 71,72/

(e) ______________________________ 73,74/

75-78/0202

---

12. HAND RESPONDENT THE SHEETS MARKED QUESTION # 12 AND PLACE PHOTOGRAPH "A" BEFORE HIM.

This question is similar to the earlier question which you filled out by yourself. This time, however, we want you to check the locations which give your impressions of these two photographs. Start with photograph "A".

REMOVE PHOTOGRAPH "A" UPON COMPLETION OF THE TASK AND PLACE PHOTOGRAPH "B" BEFORE THE RESPONDENT.

Let's take a five minute break while I set up the materials for the game which we will play after the break.

---

13. Now we are going to play a game. The game is fun and everyone seems to like it. But it is also very important. It will give you a chance to consider how you might change the area where you live to suit you better. In the game the changes you make in the area where you live are made through a series of trades, in which you get something by giving up something. Before we start trading, however, there are a few preliminary steps which we must take first.

A. PLACE ALL THE GAME CARDS IN A ROW BEFORE THE RESPONDENT SO THAT THE RESPONDENT CAN READ THE TITLES. EXAMPLE:

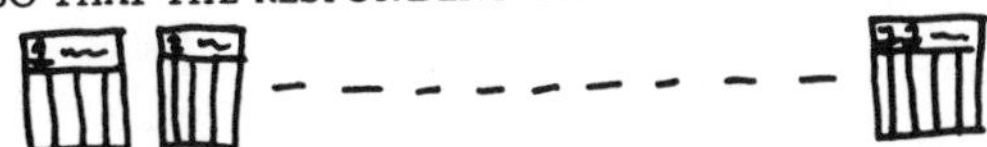

Each of these cards describes different things about where you live. Now I will give you 25 chips. Please stack the chips next to the cards in a way that represents how much you value each card. (PROBE: Are you finished?)

IF THE RESPONDENT REFUSES OR IS UNABLE TO DO THIS TASK, ASK HIM TO RANK THEM IN ORDER OF THEIR IMPORTANCE TO HIM. RECORD THE NUMBER OF CHIPS (OR RANKINGS -- 11 BEING THE MOST IMPORTANT, 1 THE LEAST) IN COLUMNS 7 AND 8 OF THE CODING FORM.

B. TAKING ONE CARD, EXPLAIN ITS FORMAT TO THE RESPONDENT.

Each of these cards has a number of descriptions for that characteristic. In everyday life we use such statements to describe the conditions of the area where we live, in terms of what we need, or what we like. I would like you to read the descriptions on each card and decide which one best describes the area where you live now, as it concerns you. (USE CARD 3 AS AN EXAMPLE) When you have decided please circle the letter in that box and tell me which one it is.

HAND THE RESPONDENTS ALL 11 CARDS -- ONE AT A TIME AND IN THEIR SERIAL ORDER -- AND RECORD THE LETTERS CALLED OUT BY THE RESPONDENT IN COLUMN 9 OF THE CODING FORM.

IF THE RESPONDENT DID NOT ASSIGN ANY CHIPS TO ANY OF THE CARDS OR HE HAS INDICATED THAT SOME OF THE CARDS ARE IRRELEVANT FOR HIM (E.G. "ACCESS TO WORK" FOR ELDERLY SUBJECTS). PUT A SLASH (/) IN THE "EXISTING (OR DESIRED) LEVEL" BOX, COLS. 9-10, OR 11-12, DISREGARD THAT CARD FOR THE REST OF THE GAME.

ASK THE RESPONDENT TO EXPLAIN WHAT HIS EXISTING LEVELS ARE FOR ITEMS 3,4,9 AND 10. RECORD HIS COMMENTS ON THE FORM BELOW.

You have checked the conditon ________ in card 3 (or 4, or 9, or 10) as existing where you are living now. Could you explain? Could you describe in specific terms?

	Description	
		1-6/
ITEM 3		7,8/ 9,10/ 11,12/
ITEM 4		13,14/ 15,16/ 17,18/
ITEM 9		19,20/ 21,22/ 23,24/
ITEM 10		25,26/ 27,28/ 29,30/

75-78/0612

Now I would like to know how satisfied you are with these different things about your place. Here is a sample satisfaction scale. POINT TO THE SATISFACTION SCALE AND EXPLAIN. This is a 7-point scale in which a 7-point score means that you are most satisfied and a 1-point score means that you are least satisfied. Now, using this card as a guide please rate the descriptions you have circled on each card. Please tell me how you would rate the description you had circled as I hand you the cards.
HAND THE RESPONDENT ALL 11 CARDS -- ONE AT A TIME AND IN THEIR SERIAL ORDER. RECORD RESPECTIVE SCORES IN COLUMN 10.

C. PLACE ALL 11 CARDS IN A ROW BEFORE THE RESPONDENT.

Now according to the rules of this game you can improve from 3 to 5 of these 11 things, but in return you will have to give up some amount of the rest. Under this condition which 5 (or less) things would you like to see improved about your place? Let's put those cards in a separate row.

SEPARATE THE CARDS IN TWO ROWS. START WITH THE ROW WHICH HAS THE CARDS TO BE IMPROVED.

By how much would you like these things to be improved, assuming your present life style, habits, work, etc., or in other words everything remaining the same. Please put a check mark on the condition that describes the improvement you would like to have on each of these cards, keeping in mind that you'll have to give up something in return, and please tell me which one it is!

RECORD THE LETTERS IN COLUMN 11 OF THE CODING FORM.

Now using this satisfaction scale as a guide, would you please rate these new descriptions?

RECORD THE RATING SCORES IN COLUMN 12.

As I have mentioned earlier according to the rules of this game you'll have to give up something in order to gain something. The way we will play this game is to take each of the cards that you wanted to improve one at a time and I will ask you how much at the most you will give up of each of the other cards (not to be improved) in exchange for that improvement, everything else remaining exactly the same. In other words, you are paying for this improvement in terms of other things that you have. I will ask you the same question for each one of the other cards, and then repeat the same process for any of the cards that you want to be improved.

BEFORE YOU BEGIN, MARK (CHECK, COLOR OR WHATEVER YOU FIND EASY TO FOLLOW) THE ITEMS TO BE CHANGED ON THE VERTICAL LIST TO THE LEFT OF THE CODING FORM. TAKE ONE CARD FROM THE PILE OF THOSE TO BE IMPROVED.

How much at the most, for example, would you be willing to give up of this item in exchange? (TAKE ONE CARD FROM THE PILE OF THOSE NOT TO BE IMPROVED AND HAND IT TO THE RESPONDENT). Please call out the letter corresponding to the appropriate box.

NOTICE THAT EACH INTERSECTION OF DIFFERENT ATTRIBUTES HAS TWO BOXES (COLUMNS 13 & 14 FOR THE 1ST ATTRIBUTE, FOR EXAMPLE). RECORD THE LETTER ON THE LEFTHAND BOX OF THE INTERSECTION OF THE ITEM TO BE CHANGED AND THE ITEM TRADED FROM THE HORIZONTAL LIST ON THE TOP OF THE CODING FORM. IF THE RESPONDENT REFUSES TO GIVE UP ANYTHING PUT A "X" AT THE INTERSECTION. CONTINUE.

Let's take another thing (TAKE THE SECOND CARD FROM THE PILE OF THOSE NOT TO BE IMPROVED). If everything else remains the same, how much at the most would you give up of this card to gain the improvement you want?

RECORD SCORE IN THE SAME FASHION. AND REPEAT PROCEDURE FOR THE REST OF CARDS IN "NO IMPROVEMENT" PILE. WHEN FINISHED, TAKE THE SECOND CARD FROM THE "IMPROVEMENT" PILE AND REPEAT PROCEDURE AS ABOVE. REPEAT FOR THE REST OF THE "IMPROVEMENT" CARDS.

ONCE ALL TRANSACTIONS ARE COMPLETE, ASK THE RESPONDENT TO RATE ALL THE NEW LEVELS HE MAY HAVE CHECKED IN THE PROCESS OF MAKING TRADE-OFFS. RECORD THE SCORES IN THE SECOND BOX OF THE INTERSECTION CORRESPONDING TO ATTRIBUTES TRADED-OFF, WHICH YOU READ FROM THE HORIZONTAL LIST.

D. NOW PLACE ALL 11 CARDS IN A ROW BEFORE THE RESPONDENT.

Once more I will give you 25 chips. Please stack the chips next to the cards in a way that represents how much you value each card. (PROBE: Are you finished?)

IF THE RESPONDENT REFUSES OR IS UNABLE TO DO THIS TASK, ASK HIM TO RANK THEM IN ORDER OF THEIR IMPORTANCE TO HIM. RECORD THE NUMBER OF CHIPS (OR RANKINGS -- 11 BEING THE MOST IMPORTANT, 1 THE LEAST) IN COLUMNS 35 AND 36 OF THE CODING FORM.

1 ACCESS TO WORK

A. Within 5 minutes walk	
B. Within 15 minutes walk	
C. Within 10 minutes drive	
D. Within 20 minutes drive	
E. Within 30 minutes drive	
F. Within 45 minutes drive	
G. Within 1 hour drive	
H. Within 15 minutes bus ride	
I. Within 30 minutes bus ride	
J. Within 1 hour bus ride	

2 ACCESS TO FRIENDS & RELATIVES, ETC

A. Within 5 minutes walk	
B. Within 15 minutes walk	
C. Within 10 minutes drive	
D. Within 20 minutes drive	
E. Within 30 minutes drive	
F. Within 45 minutes drive	
G. Within 1 hour drive	
H. Within 15 minutes bus ride	
I. Within 30 minutes bus ride	
J. Within 1 hour bus ride	

3 ADEQUACY OF DWELLING SPACE

A. Lots of extra space	
B. Some extra space	
C. Just enough space	
D. Not enough space	
E. Extremely cramped	

4 AIR QUALITY

A.	Extremely Clean	
B.	Clean	
C.	Somewhat Clean	
D.	In Between	
E.	Somewhat Smoggy	
F	Smoggy	
G	Extremely Smoggy	

5 ACCESS TO SCHOOLS FOR CHILDREN

A.	Within 5 minutes walk	
B.	Within 15 minutes walk	
C.	Within 10 minutes drive	
D.	Within 20 minutes drive	
E.	Within 30 minutes drive	
F.	Within 45 minutes drive	
G.	Within 1 hour drive	
H.	Within 15 minutes bus ride	
I.	Within 30 minutes bus ride	
J.	Within 1 hour bus ride	

6 ACCESS TO CULTURAL & ENTERTAINMENT OPPORTUNITIES

A.	Within 5 minutes walk	
B.	Within 15 minutes walk	
C.	Within 10 minutes drive	
D.	Within 20 minutes drive	
E.	Within 30 minutes drive	
F.	Within 45 minutes drive	
G.	Within 1 hour drive	
H.	Within 15 minutes bus ride	
I.	Within 30 minutes bus ride	
J.	Within 1 hour bus ride	

7 NUMBER OF HOUSEHOLDS ON AN AVERAGE BLOCK

A.	Less than 10 households per block	
B.	Between 10-25 households per block	
C.	Between 25-35 households per block	
D.	Between 35-60 households per block	
E.	Between 60-100 households per block	
F.	Between 100-200 households per block	
G.	Over 200 households per block	

8 ACCESS TO SHOPPING

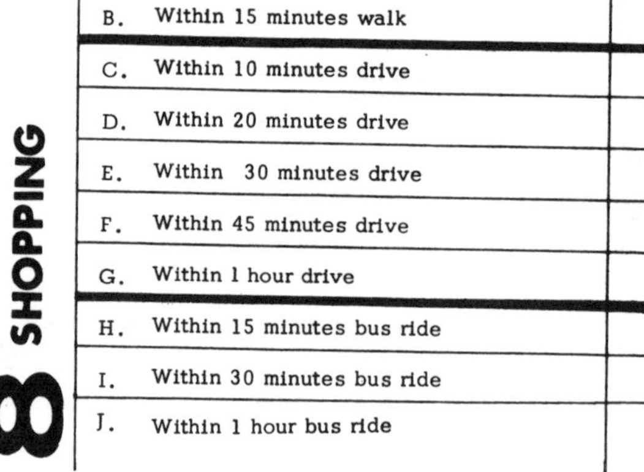

A.	Within 5 minutes walk	
B.	Within 15 minutes walk	
C.	Within 10 minutes drive	
D.	Within 20 minutes drive	
E.	Within 30 minutes drive	
F.	Within 45 minutes drive	
G.	Within 1 hour drive	
H.	Within 15 minutes bus ride	
I.	Within 30 minutes bus ride	
J.	Within 1 hour bus ride	

9 PERSONAL/PROPERTY SAFETY

A.	Extremely Safe	
B.	Safe	
C.	Somewhat Safe	
D.	In Between	
E.	Somewhat Unsafe	
F.	Unsafe	
G.	Extremely Unsafe	

10 TYPE OF PEOPLE

A.	Extremely Desirable	
B.	Desirable	
C.	Somewhat Desirable	
D.	In Between	
E.	Somewhat Undesirable	
F.	Undesirable	
G.	Extremely Undesirable	

11 ACCESS TO RECREATION (PARKS, BEACHES, ETC)

A.	Within 5 minutes walk	
B.	Within 10 minutes walk	
C.	Within 10 minutes drive	
D.	Within 20 minutes drive	
E.	Within 30 minutes drive	
F.	Within 45 minutes drive	
G.	Within 1 hour drive	
H.	Within 15 minutes bus ride	
I.	Within 30 minutes bus ride	
J.	Within 1 hour bus ride	

14. PLACE THE FOLLOWING YELLOW CARDS BEFORE THE RESPONDENT:

1. THINGS I DO

2. THINGS I DON'T DO, BUT WOULD LIKE TO DO

3. THINGS I DON'T DO

HAND RESPONDENT THE DECK OF BLUE CARDS.

We want to know some of the things that you do. To find this out I have given you a set of cards. Printed on each card is a different activity. As you read each card decide which of these categories the card best fits under, and then place the card there. Regarding the category "Things I do" please include all those things which you have done at least once in the past year.

UPON COMPLETION OF THE TASK:

1. WRAP PILE #3 AND ITS LABEL CARD WITH A RUBBER BAND.

2. SET ASIDE PILE #2 WITH ITS LABEL CARD FOR USE IN Q.17.

3. HAND PILE #1 TO THE RESPONDENT AND SET ASIDE ITS LABEL CARD.

---

15. PILE #1 FROM Q.14 IS TO BE USED FOR THIS QUESTION. PLACE THE FOLLOWING YELLOW CARDS BEFORE THE RESPONDENT:

4. ACTIVITIES I AM SATISFIED WITH

5. ACTIVITIES I AM SOMEWHAT DISSATISFIED WITH

You now have all those cards which you placed under the "Things I do" category. We now want to know which of these activities you are satisfied with and which of these activities you are dissatisfied with for reasons, such as:

a. how,
b. where,
c. with whom,
d. how often, or
e. when the activity is done.

As you read each card decide which of these categories the card best fits under, and then place the card there.

UPON COMPLETION OF THE TASK:

1. PLACE PILE #4 (EXCEPT LABEL CARD) WITH LABEL CARD "Things I do" AND SET ASIDE FOR Q. 18.

2. PLACE PILE #5 IN FRONT OF YOU FOR THE NEXT QUESTION.

---

16. PILE #5 FROM Q. 15 IS TO BE USED FOR THIS QUESTION.

Before me are all the activities which you placed under the dissatisfied category. As I read each of these cards please tell me all the reasons why you are somewhat dissatisfied with the activity. (PROBE: Are there any other reasons?)

READ EACH CARD TO THE RESPONDENT ONE AT A TIME. RECORD THE CODE NUMBER OF THE ACTIVITY CARD AND THE RESPONDENT'S REASONS ON THE CODING SHEETS BEGINNING ON PAGE E-5.

UPON COMPLETION OF THE TASK COMBINE THESE CARDS WITH THE CARDS ALREADY LABELLED "Things I do" AND SET ASIDE FOR Q. 18.

---

17. PILE #2 FROM Q. 14 IS TO BE USED FOR THIS QUESTION. PLACE PILE #2 BEFORE YOU.

Before me are all the activities which you don't do but would like to do. As I read each of these cards please tell me all the reasons why you don't do the activity. (PROBE: Are there any other reasons?)

READ EACH CARD TO THE RESPONDENT ONE AT A TIME. RECORD THE CODE NUMBER OF THE ACTIVITY CARD AND THE RESPONDENT'S REASONS ON THE CODING SHEETS BEGINNING ON PAGE E-10.

UPON COMPLETION OF THE TASK WRAP THESE CARDS AND THE LABEL CARD "Things I don't do, but would like to do" WITH A RUBBER BAND.

---

18. THE CARDS LABELLED "Things I do" ARE TO BE USED FOR THIS QUESTION.

PLACE THE FOLLOWING YELLOW CARDS BEFORE THE RESPONDENT:

1. ABOUT ONCE A YEAR
2. ABOUT 2 TO 6 TIMES A YEAR
3. ABOUT 7 TO 11 TIMES A YEAR
4. ABOUT ONCE A MONTH
5. ABOUT ONCE EVERY 2 WEEKS

(LIST CONTINUED ON NEXT PAGE)

6. ABOUT ONCE A WEEK

7. ABOUT 2 TO 6 TIMES A WEEK

8. ABOUT ONCE A DAY

HAND RESPONDENT THE CARDS LABELLED "Things I do".

On each of those cards are the things you do. We now want to know how often you do the things you do. As you read each card decide which of these categories the card best fits under, and then place the card there.

UPON COMPLETION OF THE TASK PLACE EACH LABEL CARD ON ITS RESPECTIVE PILE AND COMBINE THE PILES INTO ONE FOR USE IN THE NEXT QUESTION.

---

19. THE PILE WHICH DEVELOPED FROM Q. 18 IS TO BE USED FOR THIS QUESTION.

PLACE THE FOLLOWING YELLOW CARDS BEFORE THE RESPONDENT:

1. ABOUT 10 MINUTES

2. ABOUT 15 MINUTES

3. ABOUT A HALF HOUR

4. ABOUT 1 HOUR

5. ABOUT 1 1/2 HOURS

6. ABOUT 2 HOURS

7. ABOUT 3 HOURS

8. ABOUT 6 HOURS

We now want to know how long you spend doing the things you do. As you read each card decide which of these categories the card best fits under, and then place the card there. If you feel the time you spend on a particular activity falls half-way between two categories, place the card in the category showing less time.

1. HAND RESPONDENT THE FIRST SET OF CARDS FROM THE PILE WHICH DEVELOPED FROM THE PREVIOUS QUESTION, SETTING ASIDE THE FREQUENCY-LABEL CARD FOR THAT SET.

2. GIVE RESPONDENT TIME TO PERFORM THE TASK.

3. UPON COMPLETION OF THE TASK, PLACE EACH OF THE EIGHT DURATION-LABEL CARDS ON ITS RESPECTIVE PILE.

4. COMBINE ALL THE PILES INTO ONE AND PLACE THE FREQUENCY-LABEL CARD FOR THIS PILE ON TOP. WRAP THE ENTIRE THING WITH A RUBBER BAND.

5. PLACE A NEW SET OF EIGHT DURATION-LABEL CARDS BEFORE THE RESPONDENT.

6. GIVE THE RESPONDENT THE NEXT SET OF CARDS FROM THE PILE WHICH DEVELOPED FROM THE PREVIOUS QUESTION, SETTING ASIDE THE FREQUENCY-LABEL CARD FOR THAT SET.

7. REPEAT THE ABOVE PROCESS UNTIL ALL EIGHT SETS FROM THE PREVIOUS QUESTION HAVE USED.

END OF INTERVIEW

CODING SHEET - ELEMENTS (FILE 3)
Questions 6 & 7

Element Code	Question 6	Question 7	Element Code	Question 6	Question 7	Element Code	Question 6	Question 7
1	7/	8/	35	7/	8/	69	7/	8/
2	9/	10/	36	9/	10/	70	9/	10/
3	11/	12/	37	11/	12/	71	11/	12/
4	13/	14/	38	13/	14/	72	13/	14/
5	15/	16/	39	15/	16/	73	15/	16/
6	17/	18/	40	17/	18/	74	17/	18/
7	19/	20/	41	19/	20/	75	19/	20/
8	21/	22	42	21/	22/	76	21/	22/
9	23/	24/	43	23/	24/	77	23/	24/
10	25/	26/	44	25/	26/	78	25/	26/
11	27/	28/	45	27/	28/		75-78/0303	
12	29/	30/	46	29/	30/			
13	31/	32/	47	31/	32/			
14	33/	34/	48	33/	34/			
15	35/	36/	49	35/	36/			
16	37/	38/	50	37/	38/			
17	39/	40/	51	39/	40/			
18	41/	42/	52	41/	42/			
19	43/	44/	53	43/	44/			
20	45/	46/	54	45/	46/			
21	47/	48/	55	47/	48/			
22	49/	50/	56	49/	50/			
23	51/	52/	57	51/	52/			
24	53/	54/	58	53/	54/			
25	55/	56/	59	55/	56/			
26	57/	58/	60	57/	58/			
27	59/	60/	61	59/	60/			
28	61/	62/	62	61/	62/			
29	63/	64/	63	63/	64/			
30	65/	66/	64	65/	66/			
31	67/	68/	65	67/	68/			
32	69/	70/	66	69/	70/			
33	71/	72/	67	71/	72/			
34	73/	74/	68	73/	74/			
	75 - 78/0301			75 - 78/0302				

CODING FORM

THINGS THAT ARE BEING GIVEN UP THINGS ARE ARE BEING IMPROVED	Interview ID	No. of Chips Allocated		Existing Levels		Desired Levels		1. Access to Work		2. Access to Friends & Relatives		3. Adequacy of Dwelling space		4. Air Quality		5. Access to Schools For Children		6. Access to Cultural & Entertainment		7. Number of Households on an Average Block		8. Access to Shopping		9. Personal/Property		10. Type of People		11. Access to Recreation (Parks, Beaches, etc.)		Revised Allocation on chips		Card ID
				level	score	level	score	level	score	level	score	level	score	level	score	level	score	level	score	level	score	level	score	level	score	level	score	level	score	level	score	
Column Number	1–6/	7	8	9	10	11	12	13	14	15	16	17	18	19	20	21	22	23	24	25	26	27	28	29	30	31	32	33	34	35	36	
1. Access to Work																																75/78 0601
2. Access to Friends & Relatives, Etc.																																75/78 0602
3. Adequacy of Dwelling Space																																75/78 0603
4. Air Quality																																75/78 0604
5. Access to Schools For Children																																75/78 0605
6. Access to Cultural & Entertainment																																75/78 0606
7. Number of Households on an Average Block																																75/78 0607
8. Access to Shopping																																75/78 0608
9. Personal/Property Safety																																75/78 0609
10. Type of People																																75/78 0610
11. Access to Recreation																																75/78 0611

CODING SHEETS - ACTIVITIES (FILE 4)

Questions 14 thru 19

ACTIVITY CODE	Question 14	Question 15	Question 16	Question 17	Question 18	Question 19
Possible Codes	1,2,3,	4,5			1-8	1-8
	1-6/					
1	7/	8/	9,10/	11,12/	13/	14/
2	15/	16/	17,18/	19,20/	21/	22/
3	23/	24/	25.26/	27,28/	29/	30/
4	31/	32/	33,34/	35,36/	37/	38/
5	39/	40/	41,42/	43,44/	45 /	46/
6	47/	48/	49.50/	51,52/	53/	54/
7	55/	56/	57,58/	59,60/	61/	62/
8	63/	64/	65,66/	67,68/	69/	70/
						75-78/0401
	1-6/					
9	7/	8/	9,10/	11,12/	13/	14/
10	15/	16/	17,18/	19,20/	21/	22/
11	23/	24/	25,26/	27,28/	29	30/
12	31/	32/	33,34/	35,36/	37/	38/
13	39/	40/	41,42/	43,44/	45/	46/
14	47/	48/	49,50/	51,52/	53/	54/
15	55/	56/	57,58/	59,60/	61/	62/
16	63/	64/	65,66/	67,68/	69/	70/
						75-78/0402
	1-6/					
17	7/	8/	9,10/	11,12/	13/	14/
18	15/	16/	17,18/	19,20/	21/	22/
19	23/	24/	25,26/	27,28/	29/	30/
20	31/	32/	33,34/	35,36/	37/	38/
21	39/	40/	41,42/	43,44/	45/	46/
22	47/	48/	49,50/	51,52/	53/	54/
23	55/	56/	57,58/	59,60/	61/	62/
24	63/	64/	65,66/	67,68/	69/	70/
						75-78/0403
	1-6/					
25	7/	8/	9,10/	11,12/	13/	14/
26	15/	16/	17,18/	19,20/	21/	22/
27	23/	24	25,26/	27,28/	29/	30/
28	31/	32/	33,34/	35,36/	37/	38/
29	39/	40/	41,42/	43,44/	45/	46/
30	47/	48/	49,50/	51,52/	53/	54/
31	55/	56/	57,58/	59,60/	61/	62/
32	63/	64/	65,66/	67,68/	69/	70/
						75-78/0404
	1-6/					
33	7/	8/	9,10/	11,12/	13/	14/
34	15/	16/	17,18/	19,20/	21/	22/
35	23/	24	25,26/	27,28/	29/	30/
36	31/	32/	33,34/	35,36/	37/	38/
37	39/	40/	41,42/	43,44/	45/	46/
38	47/	48/	49,50/	51,52/	53/	54/
39	55/	56/	57,58/	59,60/	61/	62/
40	63/	64/	65,66/	67,68/	69/	70/
						75-78/0405

CODING SHEETS - ACTIVITIES (FILE 4)

Questions 14 thru 19

ACTIVITY CODE	Question 14	Question 15	Question 16	Question 17	Question 18	Question 19
Possible Codes	1,2,3	4,5,			1-8	1-8
	1 - 6/					
41	7/	8/	9,10/	11,12/	13/	14/
42	15/	16/	17,18/	19,20/	21/	22/
43	23/	24/	25,26/	27,28/	29/	30/
44	31/	32/	33,34/	35,36/	37/	38/
45	39/	40/	41,42/	43,44/	45/	46/
46	47/	48/	40,50/	51,52/	53/	54/
47	55/	56/	57,58/	59,60/	61/	62/
48	63/	64/	65,66/	67,68/	69/	70/
						75-78/0406
	1-6/					
49	7/	8/	9,10/	11,12/	13/	14/
50	15/	16/	17,18/	19,20/	21/	22/
51	23/	24/	26,26/	27,28/	29/	30/
52	31/	32/	33,34/	35,36/	37/	38/
53	39/	40/	41,42/	43,44/	45/	46/
54	47/	48/	49,50/	51,52/	53/	54/
55	55/	56/	57,58/	59,60/	61/	62/
56	63/	64/	65,66/	67,68/	69/	70/
						75-78/0407
	1-6/					
57	7/	8/	9,10/	11,12/	13/	14/
58	15/	16/	17,18/	19,20/	21/	22/
59	23/	24/	25,26/	27,28/	29/	30/
60	31/	32/	33,34/	35,36/	37/	38/
61	39/	40/	41,42/	43,44/	45/	46/
62	47/	48/	49,50/	51,52/	53/	54/
63	55/	56/	57,58/	59,60/	61/	62/
64	63/	64/	65,66/	67,68/	69/	70/
						75-78/0408
	1-6/					
65	7/	8/	9,10/	11,12/	13/	14/
66	15/	16/	17,18/	19,20/	21/	22/
67	23/	24/	25,26/	27,28/	29/	30/
68	31/	32/	33,34/	35,36/	37/	38/
69	39/	40/	41,42/	43,44/	45/	46/
70	47/	48/	49,50/	51,52/	53/	54/
71	55/	56/	57,58/	59,60/	61/	62/
72	63/	64/	65,66/	67,68/	69/	70/
						75-78/0409
	1-6/					
73	7/	8/	9,10/	11,12/	13/	14/
74	15/	16/	17,18/	19,20/	21/	22/
75	23/	24/	25,26/	27,28/	29/	30/
76	31/	32/	33,34/	35,36/	37/	38/
77	39/	40/	41,42/	43,44/	45/	46/
78	47/	48/	49,50/	51,52/	53/	54/
79	55/	56/	57,58/	59,60/	61/	62/
80	63/	64/	65,66/	67,68/	69/	70/
						75-78/0410

CODING SHEETS
Question 16

Activity Code		Reason
____	(1)	____
	(2)	____
	(3)	____
____	(1)	____
	(2)	____
	(3)	____
____	(1)	____
	(2)	____
	(3)	____
____	(1)	____
	(2)	____
	(3)	____
____	(1)	____
	(2)	____
	(3)	____
____	(1)	____
	(2)	____
	(3)	____

Appendix II

Supplementary Tables

TABLE A1. Income Ranges by Family Size for Lower-, Middle-, and Upper-Income Groups

Family size	Income groups[a]		
	Lower	Middle	Upper
1 person	$0 to $6,000	to	$19,250 or more
2 persons	$0 to $7,000	to	$20,250 or more
3 persons	$0 to $7,750	to	$21,000 or more
4 persons	$0 to $8,250	to	$21,500 or more
5 persons	$0 to $8,750	to	$22,000 or more
6 persons	$0 to $9,500	to	$22,500 or more
7 persons	$0 to $10,000	to	$23,000 or more
8 persons	$0 to $10,500	to	$23,500 or more
9 persons	$0 to $11,000	to	$24,250 or more
10 persons	$0 to $11,500	to	$24,750 or more

[a]Strictly interpreted, the upper income range for the lower-income and middle-income groups is one dollar less than shown (e.g., for 1 person $0 to $5,999, and $6,000 to $19,249), but for ease of presentation, we have simplified the ranges to the above scheme. These income ranges are based on 1970 census data. Adjusting for the increases in the Consumer Price Index, in 1980 dollars, for a typical family of four, the upper bracket of the lower income group will be $17,820, and that of the middle income group, $46,440.

TABLE A2. Breakdown of Population Groups by Stages in Family Cycle and Neighborhoods

Upper-income white	85				
		Households with children	43	Pacific Palisades	17
		Households without children	21	Bel-Air	23
		Elderly	21	Palos Verdes	26
				San Marino	19
Middle-income black	86				
		Households with children	49	Carson	36
		Households without children	24	Crenshaw	50
		Elderly	13		
Middle-income white	80				
		Households with children	43	Westchester	20
		Households without children	17	East Long Beach	20
		Elderly	20	Van Nuys	21
				Temple City	19
Middle-income Hispanic	59				
		Households with children	41	Whittier	18
		Households without children	16	Monterey Park	32
		Elderly	2	Montebello	19
Lower-income black	22				
		Households with children	17	Watts	13
		Households without children	2	Slauson	9
		Elderly	3		
Lower-income white	88				
		Households with children	37	Venice	20
		Households without children	23	Long Beach	21
				Bell Gardens	23
				Baldwin Park	24
Lower-income Hispanic	55				
		Households with children	26	Boyle Heights	16
		Households without children	14	City Terrace	16
		Elderly	15	East Los Angeles	23

Index